W9-COB-346

# FRENCH THINKING ABOUT ANIMALS

THE ANIMAL TURN

**Series Editor**
Linda Kalof

**Series Advisory Board**
Marc Bekoff, Juliet Clutton-Brock, Nigel Rothfels

*Making Animal Meaning*
Edited by Linda Kalof and Georgina M. Montgomery

*Animals as Domesticates: A World View through History*
Juliet Clutton-Brock

*Animals as Neighbors: The Past and Present of Commensal Species*
Terry O'Connor

*French Thinking about Animals*
Edited by Louisa Mackenzie and Stephanie Posthumus

# FRENCH THINKING ABOUT ANIMALS

Edited by Louisa Mackenzie and Stephanie Posthumus

Michigan State University Press
East Lansing

∞ The paper used in this publication meets the minimum requirements of ANSI/NISO Z39.48-1992 (R 1997) (Permanence of Paper).

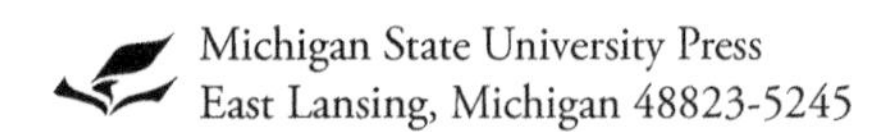

Michigan State University Press
East Lansing, Michigan 48823-5245

Printed and bound in the United States of America.

22 21 20 19 18 17 16 15     1 2 3 4 5 6 7 8 9 10

Library of Congress Control Number: 2014945801
ISBN: 978-1-61186-152-5 (cloth)
ISBN: 978-1-60917-437-8 (ebook: PDF)
ISBN: 978-1-62895-046-5 (ebook: ePub)
ISBN: 978-1-62896-046-4 (ebook: Mobi/prc)

Book design by Scribe Inc. (www.scribenet.com)
Cover design by Erin Kirk New
Cover image is *Chameleon*, seventeenth century. Oil on canvas, by Pieter Boel (1622–1674). © Musée du Louvre, Dist. RMN-Grand Palais / Photo by Gérard Blot / Art Resource, New York. Used with permission.

Michigan State University Press is a member of the Green Press Initiative and is committed to developing and encouraging ecologically responsible publishing practices. For more information about the Green Press Initiative and the use of recycled paper in book publishing, please visit www.greenpressinitiative.org.

Visit Michigan State University Press at www.msupress.org

# Contents

# Foreword

JEAN-BAPTISTE JEANGÈNE VILMER

*Translated by* STEPHANIE POSTHUMUS

Is there a French way of asking the "animal question"? First of all, we might ask: what is the animal question exactly?

Referring back to the first uses of the expression, we can respond that the animal question was principally a question of animal ethics: what moral responsibility do humans have towards non-human animals? Asked first in English, the animal question began emerging in the early nineteenth century. In an 1829 treatise on horses, John Lawrence uses the question as a section title to echo a question that he had asked forty years earlier in his "first essay on the duties of man towards those animals committed by nature to his charge."[1]

The animal question was, then, a way of denouncing the mistreatment of animals, acts of cruelty towards animals, and practices like vivisection that continue to provoke much debate today. In 1895, a publication by animal activists made the following observation: "The animal question had become a burning one during the last few years, and it was an increasingly burning question. It was a serious question of principle, and went down to the very foundation of ethics."[2]

In France, the animal question also began to appear, but with two notable differences: the first instances of the "animal question" were much later, near the turn of the twentieth century, and some of these instances did not designate an ethical problem, but instead referred to breeding practices (for example, in the *Revue de zootechnie: La revue des éleveurs*)[3] or in parliamentary debates. Asked in the Chamber of Deputies in 1921, the "animal question" referred to the "importance of the animal's pure and determined breed."[4] The differences between these two ways of asking the animal question are important.

Despite these differences, the emphasis remained on the moral significance of the animal question even as it emerged much later in the French context. This question was a social question "as gripping for the public as the question of slavery, the question of the white slave trade, and the question of child protection."[5] In 1936, Émile Germain-Sée, a French doctor, summed up the situation in the following way:

> For the sound and logical of mind, there is no animal problem. It is simply a question of knowing whether or not animals are feeling beings, are capable of pain, and if, in this respect, they have the right to be treated humanely.
>
> A scholastic, self-centered, backwards, self-seeking philosophy, having forcefully and against all evidence tried to establish and prove a theory contrary to the facts, it has become necessary to know the reasons for this claim in order to refute its arguments and establish the truth.

> In this way, there is an animal question, as complex and vast as was the question of slavery, the question of the white slave trade, and the question of child protection, not that long ago. And to have deferred for so long regulations concerning the harrowing problem of Animality will be one of the disgraces of humanity.[6]

He continues, pointing out that the animal question is a question of "the protection of animals, with the safeguarding of their rights."

The fact that the question was formulated in this way in French much later than in English was not a result of the problem not being understood. Despite what might be commonly believed, French thinking about the moral status of animals can be found in previous centuries, in the work of authors like Montaigne, Charron, Gassendi, Cyrano de Bergerac, Bossuet, Meslier, Maupertuis, Rousseau, and Voltaire. If the French tradition had been better known in the Anglophone world, we would not say that Jeremy Bentham was the first to understand that the relevant criterion for the moral consideration of animals was not rationality but suffering. Bentham formulated the question in such a way that it struck many hearts: "A full-grown horse or dog is beyond comparison a more rational, as well as a more conversable animal, than an infant of a day or a week or even a month, old. But suppose the case were otherwise, what would it avail? The question is not, Can they *reason*?, nor Can they *talk*? but, Can they *suffer*?"[7] However, twenty-five years earlier, Jean-Jacques Rousseau had arrived at the same conclusion: "If I am obligated to not harm my fellow being, it is less because he is a reasoning being than because he is a feeling being: this quality, shared by animals and humans, should give to the one, at the very least, the right to not be needlessly mistreated by the other."[8] In 1973 when Peter Singer was writing his book *Animal Liberation*, he was astounded to discover that his position, which he believed too "radical" for his friends, colleagues, and the general public, had already been defended in similar terms by Lewis Gompertz 150 years earlier.[9] He would have been even more surprised to find that the position had been taken up half a century earlier by a French thinker, Jean-Claude de la Métherie, who, two years before Bentham, expressed a clear argument for animal equality and anti-speciesism.[10]

The truth is that we have a tendency to exaggerate the "rupture" introduced by the rise of Anglophone animal ethics in the 1970s and the emergence of the "animal liberation" movement, to cite the title of Singer's well-known book. Incidentally, the expression "animal liberation" was used a few times in English well before Singer. An American text from 1839 recounts the debates of a fictional "Animal liberation society" whose aim it was to liberate wild animals in captivity.[11] The two main merits of the Anglophone movement in the 1970s, which spread quickly and now largely dominates animal ethics, is that it accelerated the movement's growth,[12] and that it developed an analytical approach that posed the problem clearly. But not one of the arguments was radically new. They can all be found in diluted form in earlier centuries.

Today, the animal question covers a much larger area and is not limited to ethics. The expression designates a set of issues—philosophical, ethical, sociological, anthropological, economic, etc.—that are raised because of the existence of non-human animals. There is, thus, not just *one* "animal question," but *many* questions about *animals* in the plural—and this plurality is important because the whole of what we call "animals" is extremely heterogeneous, including chimpanzees as well as oysters, sponges, and basic microorganisms, all of which pose different types of problems. The expansion of the area covered by the animal question is due to the considerable growth and increasing specialization of the social sciences and the humanities, whose reflections are nourished by the work of animal sciences (zoology, ethology).

Is there, then, a French way of asking this animal question? While recognizing the diversity of voices expressed in the present volume—there is certainly not *one* French way of asking the question—it is important to resist the relativist conclusion that there are as many ways of asking the question as there are people, and that nothing characterizes the French debate with respect to other ways, particularly Anglophone, of posing the animal question. I am French, have been educated in France, but have spent many years in North America and England. Having taught animal ethics in Canada, contributed to the introduction of animal ethics in France, given many lectures to different audiences, and spoken with the media on both sides of the Atlantic, I have come to the conclusion that there *are* real differences, and therefore something specific to the French way of asking the animal question.

The reasons for the lag in animal ethics in France, and for the fact that thinking about the moral status of animals is less developed in France than in many other Western countries, are threefold. First, there are the philosophical reasons. It is important to understand that humanism, which has marked the French intellectual climate for many centuries, is not just a polite defense of human rights with which everyone agrees; rather, it is a way of putting humans at the center of everything, subjugating the environment in order to fulfill the Cartesian project that humans become "as masters and possessors of nature."[13] The ecological disasters for which we are responsible, and the cruelty of which we are guilty in our treatment of animals represent the collateral damages of humanism, a philosophy that has been mobilized to justify certain cultural traditions, like the corrida (bullfighting).[14] According to the humanist view, humans and animals are like communicating vessels; so giving more consideration to the one means necessarily giving less consideration to the other. It is as if humans were not animals, as if they did not share with animals a certain number of needs—in the first place, the need to not suffer—and as if accepting this fact did not make us more *humane/human*, precisely in the figurative sense often given to this word. If animal ethics has been able to grow so quickly in the Anglophone world, it is because interest in this area has not been seen as a crime of *lèse-humanité* (against humanity).

The French intellectual climate, ingrained in humanism, is paradigmatic: the "Human" is systematically opposed to the "Animal"—a being that does not even actually exist. Derrida stated this the most clearly:

> Each time a philosopher, or anyone else, uses the "Animal" in the singular, in order to designate every living being that is not human . . . , and well, each time the subject of this statement, the "one" or this "I," utters an *asininity*. He avows without avowing it, he declares, just as a sickness is declared by way of a symptom, he offers up for diagnosis the statement "I am uttering an *asininity*." And this "I am uttering an *asininity*" should confirm not only the animality that he is disavowing but his complicit, continued, organized participation in a veritable war of the species.[15]

"Animal" is a word created by humans to distinguish themselves from all other sentient beings, to group these others into one single, abstract category to which they themselves do not belong. It is a way of denying the diversity of the living world, and, more specifically, any possible continuity between "them" and "us." It is as if a chimpanzee and a mollusk had more in common, because they are both "animals," than a chimpanzee and a human.

Admitting reluctantly that humans are animals as well—it is difficult to not do so today—in no way closes this deep divide because of how quickly we add, as if to expiate a sin, a crime of *lèse-humanité*, that humans are nevertheless "not like the other animals," that the difference is one of nature and not of degree. Humans *exist* while other animals only *live*. Humans suffer, while other animals only feel "a purely physical" pain. Humans have unique qualities that we have tried to list for many centuries: reason, consciousness, speech, morality, culture, religion, even eroticism.

The objections of scientists—zoologists, ethologists, primatologists, anthropologists—and the responses of philosophers[16] are hardly heard in this debate because of the strength of prejudices born from irrational belief, rooted in two thousand years of monotheistic religions, and maintained by humanism, and more specifically, by the interests of those who profit from animal exploitation. The "humanity" that humanism invokes is as problematic as the "animality" from which we distinguish ourselves. Evolving over time, this socially constructed concept has been used over and over again as a means of exclusion. Knowing who is human is the sinews of war, literally, because of the way we determine who deserves to be saved "in the name of humanity" (for example, in military interventions of the nineteenth century, generally to mean Christians)[17] and conversely, who can be treated inhumanely because they are not "really" human.

In 1932, Carl Schmitt underscored the dangers of this "socially constructed" humanity: "To confiscate the word humanity, to invoke and monopolize such a term probably has certain incalculable effects, such as denying the enemy the quality of being human and declaring him to be an outlaw of humanity; and a war can thereby be driven to the most extreme inhumanity."[18] The process that Schmitt describes with respect to the war of humans is precisely the one that is at work in what Derrida calls the war of the species, that is, the war humans make against other animals on the grounds that *we* are human. This paradigmatic approach, which opposes humanity and animality as mutually exclusive, is not only scientifically false and philosophically objectionable; it is first and foremost dangerous, for animals as well as for humans. It is possible as well as desirable to adopt a different paradigm.

Second, there are the cultural reasons for the lag in the French debate about the animal question. Before publishing *Animal Liberation* (1975), Singer had already understood that vegetarians—as he and his wife have been since 1971—were not welcome in France: "Unlike England, where vegetarians were uncommon but tolerated as eccentrics, in France, our requests for dishes without meat were met with open hostility. We slowly realized that it was because we were turning our back on what the French considered one of their culture's crowning glories: culinary art. It was as if we had spit on the tri-colored flag."[19] Gastronomy is an essential element of French identity that should not be underestimated in a country of breeders and hunters. There are other cultural reasons—for example, French "exceptions" like the corrida, which was banished in Catalonia, and *foie gras*, whose production and sales have been banned in California. Many French citizens support these traditions, invoking them in order to stand up proudly to "Anglo-Saxon" animal-rights theories. The same people blame "Brussels" for condemning—rightly—some of these practices. Animals pay the price for this patriotic resistance that is often accompanied by a reductive anti-Americanism—that is to say, a deep ignorance of practices and thinking outside of France.

Third, there are the political reasons that accompany and sustain the cultural ones. Lobbying that defends the interests of those who exploit animals is strong, which explains, for example, that even though only 1.8 percent of the French hunt, pro-hunting Parliament members represent 26.7 percent and 23 percent respectively of the National Assembly and the Senate[20]—members who supposedly represent the French public! This discrepancy is the result of intense lobbying efforts by hunting voters in rural departments. France is also a champion of industries related to luxury products, cosmetics, and pharmaceuticals, all of which use many animals. These other lobbying groups are careful to ensure that animal protection laws are not passed in Parliament.

These three reasons—philosophical, cultural, and political—influence how the animal question is asked in France because of the obstacles they represent. On the one hand, these obstacles have created a lag in animal ethics that has given rise to far fewer books, articles, conferences, university courses in France than on the other side of the Atlantic or on the other side of the Channel. On the other hand, these obstacles lead to the creation of new pathways as thinkers explore ways of going around and over blocked routes.

In truth, conservatism in France has not prevented the birth of what might rightly be called the "French moment." Over the last five or six years, pathways have begun to open up, thanks to the tireless work of different organizations, an increase in editorial work in this area, and more media interest in a "fashionable" subject. This "French moment" has two main orientations. First, a philosophical position has been emerging in opposition to humanism's anthropocentrism that aims to decenter the human subject. The fathers of this approach are Maurice Merleau-Ponty and Jacques Derrida. Some philosophers in this group, such as Élisabeth de Fontenay, have done much to anchor their approach in the history of philosophy, but remain too attached to anthropocentrism and speciesism, and too hostile to "Anglo-Saxon" approaches to defend anything other than *another* humanism—with half-hearted or no practical consequences in terms of new legislation or lifestyle changes, especially dietary. Other philosophers, like Florence Burgat, who has developed the strongest phenomenological approach in my opinion, go much further and so are more coherent in this respect.[21]

A second orientation of the recent "French moment" has been the introduction of animal ethics, to which I myself have contributed modestly by publishing various texts[22] and by explaining to the media that it is possible to defend animals without invoking compassion or love (since both of these create the problem of being given without being required and so do not account for the fact that animals have the *right* to be respected). As Leonard Nelson put it, animal ethics is not an attempt "to champion altruism in relation to animals. It merely reaffirms the principle of justice."[23] Presenting the animal question in this way, as an order of justice and not of charity, and defending an approach by way of interests as Singer does, has changed the image of a movement often caricatured as "sentimentalist."

The debate has begun, but the game is far from being won. Recognizing the importance that the general public now places on moral considerations, animal exploiters abuse the term "ethical" to improve their reputation and ease the consumer's conscience. Now that "ethical" meat and "ethical" fur coats are available, it becomes even more difficult to object to the need to kill for our gustatory or aesthetic pleasure. This is the perverse effect of animal welfarism: it perpetuates unjust exploitation, while at the same time reassuring the consumer that they don't need to change, because everything is "ethical."

So we must hope that the "French moment" is only just beginning, and that it will find a way to reinvent and renew itself theoretically,[24] to make itself known in other intellectual and cultural communities, starting with the Anglophone world—an effort to which the present collection of articles is contributing. We must also hope that this movement takes on the full responsibilities of its position as a way of living, so that the "animal question" is not simply an intellectual posturing, but also a social awakening that improves the lives of animals on a long-term basis and transforms us for the better.

NOTES

1. J. Lawrence, *The Horse in All His Varieties and Uses* (London: Arnold, 1829), 301. His first essay was *A Philosophical and Practical Treatise on Horses, and on the Moral Duties of Man towards the Brute Creation* (London, 1796).
2. *The Animal's Defender and Zoophilist* 14–15 (1895): 255.
3. *La revue de zootechnie: La revue des éleveurs* 4 (1925): 187.
4. *Journal officiel de la République française: Débats parlementaires*, Chamber of Deputies, first session, December 3, 1921, p. 4533.
5. Mme Even, delegate of the Department of Meuse's section of the SPCA, *Le bulletin meusien*, December 29, 1934.
6. Émile Germain-Sée, *L'animal dans la société devant la science, la philosophie, la religion* (Biarritz: Imprimerie moderne de "La Gazette," 1936), ix.
7. Jeremy Bentham, *An Introduction to the Principles of Morals and Legislation* (Oxford: Clarendon Press, 1907, 17, § 1, 4), 311, note 1.
8. Jean-Jacques Rousseau, *Discours sur l'origine et les fondements de l'inégalité parmi les hommes* (Paris: Aubier, 1973), 59.
9. Preface to Lewis Gompertz, *Moral Inquiries on the Situation of Man and of Brutes* (New York: Centaur Press, 1992), 11.
10. Jean-Claude de la Métherie, *Principes de la philosophie naturelle dans lesquels on cherche à déterminer les degrés de certitude et de probabilité des connaissances humaines*, part 1 (Geneva: n.p., 1787), 199–246; *De l'homme considéré moralement; de ses mœurs, et de celles des animaux* (Paris: Maradan, 1802), vol. 1, chap. 17: 320–322.
11. Anonymous, *Retroprogression* (Boston: James Burns, 1839), chap. 4.
12. Referring to Charles Magel's bibliography, Peter Singer notes that there were only 94 books on the moral status of animals "during the first 1970 years of the Christian era, and 240 from 1970 to 1988" and thousands more since then. In Peter Singer, *Comment vivre avec les animaux?* (Paris: Les Empêcheurs de penser en rond, 2003), 106. See also Charles Magel, *Keyguide to Information Sources in Animal Rights* (Jefferson, NC: McFarland, 1989).
13. René Descartes, *Discours de la méthode* (Leiden: Maire, 1637), vi.
14. Francis Wolff, *Philosophie de la corrida* (Paris: Fayard, 2007). See Jean-Baptiste Jeangène Vilmer, "Les sophismes de la corrida," *Revue semestrielle de Droit animalier* 2 (2010): 119–124.
15. Jacques Derrida, *L'animal que donc je suis* (Paris: Galilée, 2006), 54.
16. Florence Burgat, *Une autre existence: La condition animale* (Paris: Albin Michel, 2012).
17. Martha Finnemore, *The Purpose of Intervention: Changing Beliefs about the Use of Force* (Ithaca, NY: Cornell University Press, 2003), 66.
18. Carl Schmitt, *The Concept of the Political* (Chicago: University of Chicago Press, 1996), 54.
19. Peter Singer, preface to Jean-Baptiste Jeangène Vilmer, *Ethique animale* (Paris: PUF, 2008), 1.
20. 154 out of 577 deputies and 80 out of 348 senators are members of pro-hunting parliamentary

groups, whereas 1.2 million French people possess a hunting license (a European record) out of a total population of 65 million.

21. Florence Burgat, *Une autre existence: La condition animale* (Paris: Albin Michel, 2012).
22. Jean-Baptiste Jeangène Vilmer, *Ethique animale* (Paris: PUF, 2008), preface by Peter Singer); coedited with Hicham-Stéphane Afeissa, *Philosophie animale* (Paris: Vrin, 2010); *L'éthique animale* (Paris: PUF, 2011), and *Anthologie d'éthique animale* (Paris: PUF, 2011).
23. Leonard Nelson, *System of Ethics* (New Haven: Yale University Press, 1956), 142.
24. The interpretation that Patrick Llored champions by way of Derrida is encouraging in this sense (*Jacques Derrida: Politique et éthique de l'animalité* [Mons: Sils Maria, 2012]).

# Introduction

*French Thinking about Animals* explores the many ways in which animals signify in French history, society, and intellectual history. It brings together scholars from Belgium, Canada, France, and the United States, nine of whom wrote in French and were translated, from a variety of disciplines. They have in common that their work contributes to what might broadly—but not unproblematically, as we shall see—be considered an emerging "French animal studies." Some of our authors are leading Francophone authorities on animal histories, philosophies, rights, and behaviors, but have not to date been translated into English. Others have been translated, are about to be, or write in English. One of our goals with this volume was to start to present the diversity of French animal studies to an Anglophone readership. We also hope that this book will continue and deepen conversations between English and French animal-focused scholarship. There are of course many significant absences in terms of authors, subjects, and geographic areas; we make no claim to be exhaustive, and hope in fact that this might be the first in a number of volumes on animal meanings in global French and Francophone contexts.

As our contributors make very clear, "la question animale" has exercised widely different scholarly and activist agendas in Francophone communities for several decades (see in particular the chapters by Marie-Hélène Parizeau and Anne Simon for helpful overviews of its development).[1] But this is not to say that such work coheres around any essential cultural specificity, nor that there is really any desire for such coherence. When we asked the philosopher Isabelle Stengers—whose work certainly crosses linguistic boundaries in its inspiration and influence—to reflect on possible specificities of Francophone animal-oriented scholarship, she wondered whether the question itself might not be a red herring, describing "la Francophonie" as a "terrain of somewhat exaggerated individualities."[2] This e-mail exchange led us to call some of our own preconceptions into question, and to wonder whether the desire to seek parameters of a "French animal studies" might come more from outside, in particular from North American and British English-speaking traditions. Indeed, when we first imagined this volume, we were working with a set of assumptions, which were soon complicated: (1) that there were such entities as Anglophone animal studies and French animal studies, (2) that they could be broadly described in ways that would reveal their cultural distinctness, and (3) that they could productively be put into conversation by projects such as this volume. The ways in which these assumptions were challenged are as important as the ways in which they may have been borne out. Ultimately, we believe that the essays in this volume show a deep diversity of approaches, of engagements with animal questions and animals themselves, and are impossible to reduce to any kind of French cultural or intellectual common denominator.

The book's title itself invites the question: Is there in fact such a thing as a specifically "French" thinking on animals? How do we define its implied interlocutor, Anglophone animal studies? The latter is the preferred name in English-speaking academic contexts for a nexus of interdisciplinary research on human-animal relationships, including but not limited to (bio) ethical, philosophical, literary, historical, anthropological, and legal questions. Often, but not always, an interest in animal studies signals a commitment to an activist praxis; some scholars in the field are more explicitly aligned with theoretical reconfigurations of the human-animal divide, while still others question the very basis of the activism-theory divide. Critiques of speciesism are deployed not only to question human privileges over the non-human, but also (more recently) to question systems of oppression—racial, gendered, etc.—within humanity.[3]

The term "critical animal studies" is often used to indicate a confluence of focuses between the self-reflective process of critical thinking and theory, and consideration—often animal-rights-based—of the place of animals in the human world. As tempting as it might be to see critical animal studies as more activist than animal studies, there is within the former much lively debate between "activist" and "theoretical" approaches to the animal question. This debate certainly holds in French-speaking contexts also. While the French scholars whose work has to date gained purchase in the Anglophone academy might be characterized as theoretical (Derrida, Deleuze and Guattari, Agamben—who, although he wrote in Italian, is highly formative in French animal studies), their deployments in both French and English worlds complicate that binary. And we too believe that theory and praxis cannot be so easily opposed. For example, one of the most outspoken critics of factory farms in France, Jocelyn Porcher, has insisted that we need to pay attention to "shared suffering" between animals and the humans that work with them.[4] Giorgio Agamben in fact raises a related question in *The Open*, usually considered a densely philosophical work: "whether the well-being of a life that can no longer be recognized as either human or animal can be felt as fulfilling,"[5] a question that would resonate deeply with animal-rights groups concerned with conditions on factory farms. In this case, combining Agamben's theoretical concept of "bare life" may be a useful way to frame Porcher's empirical investigations of human and animal conditions on factory farms. This in turn could lead to policy changes that move us in the direction of what Porcher describes as a "positive and lively relationship with animals."[6] In other words, theory does not spin its wheels recursively in a closed system, only producing more theory; it is at the heart of effecting change as it affords us new concepts for arguing and implicating less oppressive relationships with animals. It "prompts detailed examination of grassroots realities . . . It can help many people understand the overall significance and fine-grained details of actual interactions with other living beings, just as it can elucidate specific social movements that address local problems with real animals."[7]

The work of Luc Ferry has also contributed to the notion that French intellectuals are not particularly concerned with animals in and of themselves, nor especially with animal rights. In his *Nouvel ordre écologique* (1992), translated fairly quickly into English, Ferry argues that the ideological history of deep ecology, including calls for animal rights, is entwined with Fascist ideals and a devaluing of humanity.[8] This does not mean, however, that Ferry espouses a Cartesian indifference to animal suffering. In the *Nouvel ordre écologique* itself, he states that animal suffering should not be imposed where not necessary, while strongly critiquing Peter Singer's notion of speciesism for equating animal with human dignity. More surprisingly, perhaps, Ferry was one of twenty-four prominent French intellectuals to sign a manifesto, published on October 23, 2013, by the animal-rights group *Fondation 30 Millions d'Amis*, in favor of the changing of France's civil code to grant that animals are "alive and feeling" rather than legal things or possessions.[9] This manifesto, signed by respected intellectuals from diverse disciplines (including Florence Burgat, a contributor to this collection), is proof of the mainstreaming of concern for the status of real animals in France. The philosopher Daniel Salvatore Schiffer opined, in an editorial in the mainstream *Le Nouvel Observateur* published three days later, that France had too long been held back in its attitude towards animals by the legacy of Descartes, and that the manifesto represented a step in the direction of a "civilization in the true sense of the term."[10] It is hard to imagine that such a public statement, in the United States at least, would be received as anything but fringe activism, but one of the lessons of French animal studies is perhaps that taking responsibility for the well-being of animals need not lead to a devaluing of the human but rather to its evolution. This is one of the premises of Élisabeth de Fontenay's important intervention *Sans offenser le genre humain*, which takes to task the (explicit or implicit) devaluing of humanity by thinkers such as Paola Cavalieri and Peter Singer.[11] Without abandoning a commitment to and curiosity about humanity, then, the French "animal moment" today is in fact highly ethically engaged with animal rights and the specificity of animal experiences and histories (see especially Florence Burgat's and Éric Baratay's chapters here).

Until recently, Anglophone readers may not have been aware of the richness and diversity of thinking on animals in the French-speaking world. And while French scholars are generally more likely to have read sources in English than vice versa, they were mutually slow on the uptake for a while. This is changing rapidly, and a consensus is emerging on the interest of a broader internationalization of thinking on animals. As Anne Simon's contribution to this volume makes clear, there are many emerging English-French collaborations, and a forthcoming volume of *French History* coedited by Peter Sahlins, also a contributor here, is dedicated to "Animals in French History." Élisabeth de Fontenay, whose work was foundational to the reorientation of French history to a more animal-centered approach, has recently been translated into English, and several of our contributors (Dominique Lestel, Vinciane Despret, Éric Baratay) have published in English and/or have translations forthcoming.

And yet, while any monolithic Frenchness of content and subject matter is of course illusory, we believe it to be an important illusion. As scholars, in both English- and French-speaking contexts, we continue to invest in the "Frenchness" of intellectual traditions as a category that matters.[12] Therefore, it does matter, but only inasmuch as we are willing to put our own investment in these categories under the microscope. Cultural differences are identified and prioritized depending on the observer's position, which is always within and never outside of a specific sociohistorical context. As ethnologists and anthropologists have convincingly argued, no neutral, objective position exists from which an observer can simply note the differences between two cultures; observing is always done from within a cultural context. In the case of the two editors of this collection, we are scholars of French literature who do our research in North America and so are keen on creating bridges between these two cultural, literary, and linguistic communities. At the same time, we are aware of the dangers of asserting culture as some bounded, homogenous, and unchanging unit that supposedly determines how an individual behaves. Such a concept of culture does not fit with our experiences of cultural exchanges, dialogues, movements. So if we argue for a sense of culture, it is with this more fluid notion in mind, following the lead of comparative literature scholars such as Homi Bhabha and Gayatri Spivak, who argue for a model of culture where hybridity, change, and cross-cultural influences are the norm. In short, culture matters because it is through the experience of cultural difference that we begin to see culture as contested and transitory.

Language is an important departure point for the exploration of cultural differences, because of how closely bound together they are. This is why we have attempted whenever possible in the translations of the nine original chapters that were submitted in French to highlight the difficulties encountered when trying to express ideas that had no literal equivalents in English (see Delannoy's use of the word "vivant," Lestel's "posture," or Despret's "*à partir de*"). Rather than gloss over these difficulties, we have seen them as opportunities for creating fruitful points of encounter where cultural and intellectual exchange can happen. These points of encounter also remind us of the linguistic frameworks we use to give meaning to the world. One of the objectives of translation studies and comparative studies is to make us more aware of the linguistic and cultural idioms that give form and structure to our thinking. We have attempted to follow this model in our collection, avoiding foregone conclusions and reductive assumptions. We hope this will serve as an example for further contributions to the cross-cultural conversations that are beginning to take place between animal studies scholars working to enrich human understandings of the non-human and vice versa.

There are gaps and absences in this collection for which we as editors take full responsibility. Readers will certainly notice the absence of contributions from widely read French thinkers who have already profoundly shaped Anglophone thinking on animals and the non-human world in general, such as Michel Serres, Bruno Latour, or Isabelle Stengers. In France, the work of Élisabeth de Fontenay has been foundational, and scholars like Jean Christophe Bailly, Georges Chapouthier, Jean-Claude Nouët, Jean-Paul Engélibert, Alain Romestaing, Audrey Lasserre, Alain Schaffner, Pascal Picq are but a few of many scholars and activists at the forefront of the current French "animal moment" who are also gaining visibility in international collaborations and/or translations. There are also many scholars publishing and working in English-speaking institutions on French animal history and culture—to name but a few, Adeline Rother, Anne Mairesse, Lucile Desblache, and Eliane DalMolin. The inherent limit, however, of a one-volume project inevitably means that it makes no claim to be exhaustive. More problematic, however, is the reduction of "Francophone" thinking on animals to essentially exclude what is produced outside France, usually in countries formerly colonized by France. The work being done by scholars such as Bénédicte Boisseron, Neel Ahuja, Achille Mbembe, and Colin Dayan is revealing the serious limitations and blind spots inherent in reducing thinking about, and representations of, animals to those produced in First World contexts. However, it was important for us not to be tokenistic about this admittedly essential move, and we decided rather to indicate the need for future volumes that would be entirely or mostly framed by postcoloniality. We would like to think that this decision was because we take seriously the need to articulate animal studies with postcolonial scholarship (i.e., we did not want to fall into the trap of tacking on one or two "Francophone" chapters as token gestures towards inclusivity in a volume that would still be overwhelmingly French), but it is equally true that it reveals our own positions as citizens and scholars of First World countries.

Our contributors are all excellent examples of the interdisciplinarity that is required of us if we are to take the animal question seriously. If they do choose to focus on one particular disciplinary approach, it is always with a thoughtfulness about the specificity of what this particular discipline can add to an aggregate of knowledge about animals. We have divided the book into four sections: animal histories, animal philosophies and representations, human-animal intimacies, and animal-oriented environmentalisms. There is a great deal of overlap between them (for example, two of the chapters on animal intimacies could equally well be in the section on philosophies, given their sustained engagement with the philosopher Jacques Derrida), and we do not mean to imply any rigorous distinction between sections.

Éric Baratay starts the first section, "Animal Histories," by arguing for greater articulation between history and ethology in order to create a truly *animal* history, one that takes into account what Jean-Christophe Bailly has termed "the animal side." While his chapter is primarily a position piece, he also provides examples of the kind of history he has in mind, in particular the experience of dogs and horses during World War I (the subject of his forthcoming book). Particular animal histories are fleshed out in the next two chapters. Peter Sahlins traces the fates of three chameleons in the academic and salon cultures of seventeenth-century Paris, in order to show how the narratives written about these particular animals articulated literary and scientific discourses, thus offering a form of resistance to the dominant Cartesian model of animals as beast-machines incapable of thought and "passions." Walter Putnam invites us to consider the experiences of the captive animals of the "living museum" in the 1931 colonial exposition in Versailles, showing how exploitation and exoticization of people and animals intersected in the French colonial projects, and the extent to which attitudes to animals very often reflect attitudes to human subalterns.

The second section, "Animal Philosophies and Representations," starts with Florence Burgat's call for an animal philosophy, echoing Baratay's animal history in the first section. Burgat argues that if philosophy truly takes into account the multiple specificities of animals, as opposed to a generalized concept of "the animal," it can provide a way out of the impasse of dualism versus continuism (humans are either not-animal or another-animal). Philosophy must be inflected with an etho-phenomenological apprehension of lived animal experiences. Dominique Lestel also draws on phenomenology, but opts for an ultra-continuist position as he argues that humanity is constructed in and through an animal *texture* that goes far beyond simple cohabitation. The philosopher must "deconstruct and reconstruct" the animal sciences in order to reveal the relational and biological interdependency (like the fingers of a hand) between human and non-human—and how they might be moving together towards a technologically determined post-animality. Turning to literary representation, Anne Simon provides a wide-ranging overview of work being done in both French and English contexts on animals in/and literature. She argues for the importance of comparative consideration of literary animal studies in both French and English, which reveals critical differences and continuities—for example, the very different reception in the United States and France of area "studies" in general, the different relations to concepts of nature and ecocriticism, and the role of theory.

The third section is on intimacies: between human and animal, but also the intimacies between philosophical traditions that arise from thinking *with* and *from* animals, as Vinciane Despret argues in her chapter. Starting with Donna Haraway's reading of Derrida, which led her to Derrida herself, Despret meditates on the meaning of inheritance. Through the "parable of the twelfth camel," which is about the need to transform the terms of that which we inherit, she subtly shows how her own philosophical heritage—Derrida, itself transformed by Haraway—led her to a different, less distanced apprehension of animal experience. The second chapter, by Carla Freccero, argues that Derrida's thinking about the non-human, via his particular cat, adds to queer theory inasmuch as it invites ethical intimacy between human and animal, thus decentralizing humanism (and phallocentrism—it is a female cat). While Derrida's cat is, as he insists, a particular animal and not an emblem of "the" generalized animal, it is also an opening up to writing, language, and to intertextual feline presences; he thus questions whether humans can in fact ever apprehend animals outside our own inscriptions. In the third chapter of this section, the legal scholar Marcela Iacub takes up sexual intimacy between humans and animals. She presents the case of a French man who, having had intercourse with his pony, was convicted of "serious abuse" of an animal. Since it was not at all clear that the pony suffered any pain, she argues, and since large-scale cruelty is inflicted on animals every day without the state taking a stance (foie gras, bullfighting, to say nothing of individual cases of torture that receive much lighter penalties), anti-bestiality laws in France serve a human regime of shaming of certain sexual passions more than they serve the interests of animals, collectively or individually.

The fourth section, "Animals and Environment," investigates the ways in which thinking about animal issues has intersected or not with more general environmental concerns and ethics. Isabelle Delannoy holds to a larger definition of the *vivant*, insisting that animals are one group within a larger living world whose vitality and intelligence need to be further recognized if we are going to move towards a fuller ecological awareness. She asserts that we are already embarking on this path, as a global concern for the environment is growing that goes beyond any one country or culture. Nathalie Blanc offers a more specific example of how such awareness is being played out with respect to human-animal relationships in urban environments. Blanc identifies the categories of nature as wanted or unwanted to explore different understandings and representations of animals, citing her studies combining geography and sociology and specific interviews with urban dwellers. Finally, she asks how the creation of greenbelts in French cities might continue to influence human-animal movements and interactions with respect to shared common spaces. Marie-Hélène Parizeau outlines a larger historical and cultural framework for the debates that have taken place between animal studies and environmentalism. She asserts that the schism between these two areas of critical studies is more characteristic of the Anglophone community than the French-speaking one. Her point is nicely illustrated by Delannoy and Blanc's articles, which do not set animal rights and environmentalism against each other, but instead explore their points of contact. Referencing her own work on bioethics, Parizeau suggests that the relational, hybrid nature of French thinking about animals might be a more useful model for responding to the sorts of animal chimera and clones currently being created by contemporary science and technology.

Parizeau's chapter, which invites us to reflect on the future of this extremely diverse area of study, nicely closes the collection. While the thoughtful contributions to the current collection fall largely within the purview of the humanities disciplines (Delannoy, Blanc, and Iacub being the notable exceptions), they nevertheless push against the limits of the humanities as having to do exclusively with the human. We might then ask: What future directions will French thinking about animals take as science and technology continue creating new forms of *le vivant*, as global warming causes more and more species to go extinct, as we as humans evolve to become something other than human? Will the animal-human distinction eventually disappear, and how will disciplines like the humanities then respond to such changes? Will the humanities become something like a non-humanities or an in-humanities? Without predicting the future of the humanities, we can assert with confidence that the interdisciplinarity and the internationalism that are at the heart of animal studies will continue to transform the ways in which the human and the non-human are defined so that the human sciences are seen to coalesce in important and meaningful ways with the natural and social sciences.

## NOTES

1. A good introduction to the range of engagements with animals among French scholars is *La question animale: Entre science, littérature et philosophie*, ed. Jean-Paul Engélibert, Lucie Campos, Catherine Coquio, and Georges Chapouthier (Rennes: Presses Universitaires de Rennes, 2011).
2. Personal e-mail to Louisa Mackenzie, March 23, 2012.
3. A comprehensive introductory footnote to Anglophone animal studies would take several pages; we refer our readers to Paul Waldau's *Animal Studies: An Introduction* (Oxford: Oxford University Press, 2013), which offers a richly intersectional and comparative review, and the ongoing online bibliography by Linda Kalof, Seven Mattes, and Amy Fitzgerald, part of the Animal Studies Program at Michigan State: http://www.animalstudies.msu.edu/bibliography.php.
4. Among her many books in French on the need to rethink the relation between farm animals and those who raise and slaughter them, see for example *Éleveurs et animaux, réinventer le lien*, with preface by Boris Cyrulnik (Paris: PUF, 2002). She is less well-known to English speakers, but see Jocelyn Porcher, "The Relationship between Workers and Animals in the Pork Industry: A Shared Suffering," *Journal of Agricultural Environmental Ethics* 24 (2011): 3–17.
5. Porcher, "The Relationship," 77.
6. Porcher, "The Relationship," 14. Interestingly—and this might indeed constitute a Francophone specificity—many opponents of factory farming in France, including Porcher and Dominique Lestel (one of our contributors) go to some lengths to distance themselves from vegetarian activists. See Lestel, *Apologie du carnivore* (Paris: Fayard, 2011). For a plea for vegetarianism, see fellow contributor Marcela Iacub's *Confessions d'une mangeuse de viande: Pourquoi je ne suis plus carnivore* (Paris: Fayard, 2011), as well as various articles by Enrique Utria. Lestel and Iacub debated each other publicly in the Centre Pompidou lecture series Hors Pistes in 2012 "Homme = carnivore? Une rencontre entre Marcela Iacub et Dominique Lestel," which can be viewed online at http://www.centrepompidou.fr/cpv/ressource.action?param.id=FR_R-9a7c763471b55cd53ab7cc351968dd¶m.idSource=FR_E-18c1df3c1dcc684bafad929b4898161.
7. Waldau, *Animal Studies*, 189.
8. Luc Ferry, *Le nouvel ordre écologique: L'arbre, l'animal, et l'homme* (Paris: Grasset, 1992), translated as *The New Ecological Order* by Carol Volk (Chicago: University of Chicago Press, 1995).
9. The manifesto can be viewed online at http://www.30millionsdamis.fr/fileadmin/user_upload/actu/10-2013/Manifeste.pdf. The civil code, which currently grants to animals the same status as nonliving possessions, is considered in Marcela Iacub's contribution to this volume. The manifesto calls for a new legal category between humans and things, one specific to animals.
10. Daniel Salvatore Schiffer, "Droit des animaux: Le changement de leur statut juridique est un enjeu de civilisation," *Le Nouvel Observateur*, October 26, 2013.
11. Élisabeth de Fontenay, *Sans offenser le genre humain: Réflexions sur la cause animale* (Paris: Albin Michel, 2008). English translation by Will Bishop, *Without Offending Humans: A Critique of Animal Rights* (Minneapolis: University of Minnesota Press, 2012). De Fontenay is also a signatory of the manifesto.
12. See, for example, *Time* magazine's front-page article proclaiming "The Death of French Culture" (December 3, 2007), and the reactions to it in both France and the United States.

# PART 1

# Animal Histories

# Building an *Animal* History

ÉRIC BARATAY

*Translated by* STEPHANIE POSTHUMUS

Since the field of animal studies has opened up, the human and social sciences, in North America and in Europe, have developed an almost exclusive interest in the human side of this subject, examining human uses, practices, and most particularly human representations of animals, in part because of a certain scholarly infatuation with cultural studies since the 1980s.[1] Having used these approaches myself many times, I now feel they are insufficient because they have created and maintained a blind spot at their center—that of animals as feeling, acting, responding beings, who have their own initiatives and reactions. Scholars have had much to say about humans and very little to say about animals, who remain absent or are transformed into simple pretexts, pure objects on which human representations, knowledge, practices are exercised without consequence. In this sense, the history of animals that has developed over the last thirty years is in reality a *human* history of animals, where these latter have very little place as real beings.

## LOOKING AT REAL ANIMALS

We must move away from this approach rooted in a Western cultural worldview that has impoverished the dialectical theme of humans *and* animals, reducing it to a field with one magnetic pole (humans) and a single directional pull (humans towards animals), thus forgetting or dismissing much of its reality and complexity. We must look more closely at the influence of animals in their relationships with humans, at their role as actual actors, in light of ethology's growing insistence—at least for certain species and an increasing number of them—on the behaviors of each animal as actor, individual, and even person; on the cognitive capacities of animal individuals; and on the sociability and cultures of animal groups—and thus revealing the inadequacies of purely human approaches. Similarly, historical documents show, when this information is not rejected as anecdotal, that humans have seen or foreseen and assessed animal interests and have reacted, acted, and imagined as a result.

We must leave the human side, moving to the animal side,[2] in order to better understand human/animal relationships, but also in order to better know these living actor-beings who deserve to be studied in and of themselves. This means that the definition of history must be broadened, abandoning the too restricted definition of "a science of humans in time,"[3] in which many historians have become entrenched. This definition is not inviolable; it has been historically constructed, from Fustel de Coulanges to Bloch, with two events of particular importance: first, the formation of the human sciences as a means to study the human independently of the natural sciences that previously had had a certain monopoly on knowledge; and second, the broadening of the human sciences, during the first third of the twentieth century, to include the study of all aspects of the human and not just those related to the political. It is now time to expand this definition of history to mean the "science of all living beings in time" and to take into account these living beings' evolutions, at the very least those that have been recorded in diverse historical documents and that could be appropriately studied by a historian versed in the field.

At the same time, we must not so much abandon as go beyond the cultural approach that tends to reduce the human and social sciences to an exercise in deconstruction and close examination of social discourses, and thus consider representations to be the only observable reality. This work is necessary; but the success of cultural approaches has transformed an essential preliminary step into an ultimate finality. We must once again be searching for realities using the concept of "situated knowledges"[4] to build knowledge that is neither ignorant of, nor taken in by its context of elaboration. We need to apply this to the diverse human actors who have used, become close to, and observed animals, and who have become witnesses to animals in varying degrees using observation and representation. We need to take into account the conditions under which these discourses were produced so that when we gather, test, and critique information that is partial—in both senses, incomplete and biased—we arrive at some sense of that reality.

We must also abandon the culturally constructed Western notion of animals as passive beings and see them instead as feeling, responding, adapting, and suffering. In other words, we need to start with the hypothesis that animals are not only actors that influence humans, but that they are also individuals with their own specific set of characteristics; they are even people with their own behaviors; in short, they are subjects. These ideas are no longer taboo[5] and should be tested in the field while leaving room for some flexibility in how the definitions are used. We must refrain from starting with (too-well) defined concepts whose reality we hope to prove, because then we simply configure these concepts according to the form we know best, that is, the human form, or more precisely the European human form at a given time, and once again we fall into the trap of ethnocentrism and anthropocentrism. We must realize that our concepts are always situated: in time, as historians show us; in space, as ethnologists point out;[6] and among living beings, as ethologists are beginning to demonstrate.[7]

Western culture has defined the subject as thinking, self-conscious, and having recourse to conscious choices and strategies, all the while forgetting that this definition—which it takes as *the* definition—is in fact a situated, inferred version of the human. Moreover, this underlying portrait includes a set of philosophical implications that place humanity at the top as absolute reference, just as the Western world placed itself at the top in the past. When one clings to this definition while observing animals, one uses a discourse of domination as a tool of investigation, arriving at the already drawn conclusion that there are no subjects among animals. It is when more supple definitions are adopted that one can envisage the concept of animals as subjects, or come to a conclusion even if not all the parameters are met. We must remember that we have just barely begun to search for these parameters in the animal world; if we find that these parameters lack some consistency, it may be that we need to consider a greater plurality of meanings.

Experimenting with key concepts does not mean falling into the trap of anthropomorphism, just as attributing flexibility and suppleness to concepts under investigation does not mean slipping into vague impressionism. What such an approach entails is a form of critical anthropomorphism that watches with curiosity, asks difficult questions, tries out critical concepts, observes without prejudice, and avoids an already conclusive anthropomorphism that foists humanity on animalities and thus denies their specificities. It also entails being as open as possible to the potential capabilities of animals who we still do not know very well. Finally, this means seeing the diverse expressions of different faculties in order to adopt wider definitions of them. This is already being done for physical abilities (we know that many species do not see the world as we do, but we do not deduce from this that they cannot see), but we remain reticent when it comes to doing the same for mental abilities because these are what allow us to value ourselves over animals.

This is not a question of mixing up all living beings, but rather it is a question of appreciating the diversity of all and the richness of each one. This means abandoning the shallow, puerile, distorted dualism that opposes humans to animals and in which philosophies and religions have trapped us for the last 2,500 years. First, this dualism is shallow because it opposes a concrete species, the human, to a concept, *the animal*, that does not exist in the fields nor in the streets, and that is nothing more than a category masking the reality of a multiplicity of species that are each very different. Second, this dualism is puerile because it poses the question of a difference between a reality and a concept not in order to better know a diversity of animals, but rather to value humans over all other species. Third, this dualism is distorted because it has established differences on a set of ungrounded beliefs; we still understand animals very poorly (and we hardly seem to want to know them better, often preferring our own convenient stereotypes). We must abandon anthropocentrism that defines concepts using humanity as a yardstick, that refuses to see or accept the side of animals, and that thus forecloses discovery before even getting started.

To build an *animal* history, history needs the help of other sciences: (1) ecology, in order to reconstruct the milieus in question and determine their influence on behavior; and (2) ethology, in order to interpret these behaviors. This latter science raises the problem of choice because there is a large divergence between successive or concurrent schools of ethology, with each arriving at a slightly different picture of animals.[8] Moreover, much ethological work is done at the level of species, and as such does not integrate social and individual dimensions. So we must draw on the most innovative work of cognitive ethology that grants the highest mental functions to animals, and in particular the functions of thinking on specific levels, social and individual. This dialogue with the so-called natural sciences should not be any more difficult for historians than conversations with the science of economics, demographics, sociology, etc., that have been happening since the middle of the twentieth century. It simply continues the tradition of opening history up to other sciences, and cannot be dismissed as less legitimate just because it moves away from the human and social sciences. Such an objection again restricts the definition of history to a human one, and echoes the attitude of historians in the early twentieth century who refused a social-science history on the grounds that history could only be political.

In the search for a complex animal history that includes individualized, personalized animals, actors, and subjects, in groups with greater sociability, it is clear that strictly biological interpretations do not suffice for understanding such complex behavior, even according to some ethologists. The historian thus needs the interdisciplinary experience of those who work at the crossroads of the natural sciences and the human and social sciences, those who already deploy concepts such as group, family, sociability, action, individual, person, subject, and who have developed appropriate methods for analysis. The development of transdisciplinary approaches, methodologies, and studies will give rise to an animal science that combines or even unifies the strengths of zoological science (which should be renamed as the biological sciences) with the strengths of the human and social sciences (which should be renamed as the cultural sciences), and that goes beyond merely animal objects for the first group and merely human for the second. But this cooperation, which should lead to a complete reconstruction of all the disciplines, is still in its beginning stages, with ethologists, ethnologists, psychologists, and historians trying out similar types of observation and analysis in different animal fields.[9]

In the meantime, the history-ethology combination is essential for examining available archives, which are most often human-focused. It may seem paradoxical to use such archives to search for animal acts and gestures, especially since questions can be raised about their reliability and about their partial, biased, sporadic character. Humans tend to be interested in just a few species, breeds, or individuals, and in just a few aspects that they have noted only sporadically, according to what they see or want to see, often interpreting and misrepresenting according to the ideas, interests, and certainties of one species of one society at one time. But these problems arise even when doing human history, for which historians must often work through intermediaries. For example, the majority of historical sources about peasants are written by people with the social prestige of the ruling class, but this has not kept historians from using them to write about rural history. In the case of animal history, the difficulty is obviously greater but not radically different. In addition, it is always by way of humans, their writings, their photographs, their films, that animals today bear witness.

We must make do with what we have. This does not mean that we do not need to sort through and select, looking for those documents written by humans who were interested in real acts and gestures, who observed and recorded without making these acts and gestures disappear under the weight of their own subjectivity. Of course, these observations will not all be equally accurate, the records not all equally well written, which means it is necessary to bring these documents together with present-day ethology's most fruitful hypotheses or knowledge. This does not mean validating the former using the latter; the temptation would be to reject or misinterpret the historical documents because of the prejudices of a time period or a way of thinking. We need instead to bring together situated knowledges and ways of seeing: those of our contemporaries, who may be blind to certain details, just as they can see what others can no longer see, or do not want to see; and those of present-day naturalists, who understand more than in the past, but who may also neglect certain details. Observations from the past must not be rejected as unusable, and recorded acts and gestures should not be reduced to the level of unusable anecdotes (although laboratory ethology did just this for a long time). Instead, these records should be seen as usable anecdotes, data from a terrain of observation situated in the past, in the same way that contemporary psychologists and cognitive ethologists are now using present-day anecdotes.[10]

In sum, moving to the animal side means standing next to animals,[11] feeling empathy so as to contest nothing in advance, adopting their geographical point of view, understanding what they feel, undergo, how they act and respond; it means trying to take on their psychological viewpoint to see what they see and to feel what they feel. Given these objectives, all more or less attainable depending on the animal being studied, what *animal* history can we build?

## HISTORY AS LIVED ANIMAL EXPERIENCE

The simplest animal history would start by inscribing animals in important human historical events, because relevant documents abound, and it is possible to see how animals lived these events in body and spirit. This is what I have tried to do for the twentieth and twenty-first centuries, during which animals have been fairly well observed, listened to, and written about.[12] For example, it is possible to develop a history of the lived experience of horses in mines during the industrial "revolution" by using the accounts of engineers, veterinarians, and miners to reconstitute itineraries and develop the individual animal's adaptations and resistances. We can thus demonstrate that these are truly animals who act and react, compelling humans to consider them (see figure 1).

A similar approach could be used to arrive at a history of dairy cows, who have been at the center of the agricultural "revolution." Documents written by veterinarians and those involved in zootechnics can be used to trace the animals' initial resistance around the eighteenth century to the shifts in dairy farming that used brutal and traumatizing separation of cows from calves, and then to follow the animals' slow adaptation, through training and selection, as the relations between cows, and cows' relations with breeders, changed with the adoption of different modes of breeding. Another possible example would be the history of animals swept up in the First World War. The records of veterinarians and soldiers can be used to understand the distress of conscripted horses and dogs separated from their fellow creatures, owners, and milieus; their fright during long trips; the tensions between these animals and unfamiliar riders and trainers during training exercises; the fear felt at the front when faced with new sounds and smells and the violence suffered by soldiers; the complicity shared with other animals and

*Figure 1.* MINING HORSE TRANSPORTING WOOD IN A MINE SHAFT, LENS, FRANCE, 1906, C. COLL. YVES PAQUETTE.

*Figure 2.* Dog with machine gun cart, English soldiers, WWI, 1914. (Source: Bibliothèque nationale de France, used with permission)

humans; the fatigue and pain caused by the work; the suffering brought on by injury; the agony and finally the moment of death or discharge (see figure 2).

Let's take a moment to look more closely at this last example of animals in the First World War in order to underline the role of ethology in the development of such a history, and also to outline what history can bring to ethology.[13] The work of ethologists aids in the comprehension and interpretation of historical descriptions; for example, one can come to a conclusion about the stress felt by horses because of the behavior recorded by veterinarians during animal conscriptions in Europe and animal auctions in North America. In other words, ethology can be used to better analyze the accounts of witnesses. Moreover, it can reveal some of the aspects that were ignored and that gave rise to misconceptions or negligence on the part of soldiers at that time. It can also reveal some of the reasons for mutual misunderstanding, such as why conscripted horses refused to get into railroad cars despite the fact that the men tried over and over again, seven or eight times, yelling and even hitting; because of their panoramic vision, the horses believed they were being pushed into an obstacle while the rest of their optical field was open. One could also cite the high mortality rate of horses in the summer of 1914 due to rapid dehydration following hard work and overwhelming heat, a form of dehydration that did not set off a strong enough desire to drink, with the result that the horses did not send signals alerting the men to their need (see figure 3).

Conversely, history can make an important contribution to ethology by providing the data needed to step back from contemporary animal situations. For example, historical documents may reveal the behavioral differences between barely domesticated horses from the American or Argentinean planes and well-domesticated European horses, or the difficulties of training conscripted dogs from kennels—who were not at the time abandoned dogs, but rather

*Figure 3.* HORSE HOSPITAL, SITE OF DISINFECTION, WWI, 1916. (SOURCE: BIBLIOTHÈQUE NATIONALE DE FRANCE, USED WITH PERMISSION)

stray dogs who refused to cooperate—compared to dogs studied by contemporary ethologists: domestic dogs who have been quickly transformed into the universal, timeless, "natural" dog. Historical documents also show the inability of homing pigeons to fly at night because they were not trained to use their magnetic compass, which suggests that the use of such a compass is not innate—as contemporary ethologists continue to maintain—but rather acquired.[14]

It should now be clear that when using these approaches and methods, we construct a history that is quite different from the human version of the phenomena and events in question; it does not insist on the same points, and so allows for a different way of seeing and understanding the interaction between humans and animals, offering a more complex view of these interactions and relationships based on multiple types of animal sociability and individuality.

## ANIMAL BIOGRAPHIES

The importance of these animal individualities requires that we construct a second history, one of individuals, by way of animal biographies. The desire to write an animal's story from the animal point of view is an old one that can be found in the work of authors such as the

Comtesse de Ségur's *Mémoires d'un âne* (Memoirs of a Donkey). But this often leads to the dead end of anthropomorphism, with some notable exceptions, such as Virginia Woolf's re-creation of the lived experience of the dog Flush using correspondence from his owners. Given this danger, the few historians, professional or amateur, who have attempted to tell the story of well-known animals have preferred to do so from the human side, outlining the intellectual, political, social, or artistic influences of these animals.[15] Nevertheless, we should not stop there; we must pose the question from the animal side and try to give an account of how an animal lives, suffers, relates, and feels at a given moment or over the course of its life. The question of the personal lived experience of a particular actor is now legitimate in the human and social sciences, and helps us better to understand, feel, and thus experience to some extent the way an individual lived an event, a phenomenon, or a time period, and to understand her reactions that may in turn have influenced the surrounding context. This should also apply to individual animals.

With the issue of available archives being of primordial importance to historians, the best-known animals constitute the most available cases since they were often the subject of a large number of accounts and reports. Conversely, anonymous animals must be accessed indirectly through the study of their communities. With these documents, we must mobilize the method of reconstruction that has become standard in paleontology, prehistory, and archeology, when only a few remains or clues are available to re-create human lives, gestures, and edifices. These reconstructions are obviously hypothetical, but they are carefully weighed and verified so that they become indispensable for seeing, understanding, and developing research questions and answers. This can been seen, for example, with respect to the First World War, when museums reconstructed trenches that were not originals, but did allow for a concrete understanding of a reality that had disappeared, and thus raised some new questions; or when historians reconstructed battles in order to better grasp and understand the lived experience of soldiers. This method should be adopted for animal biographies, even if the reconstitutions are necessarily partial because they only include aspects recorded in documents that are necessarily biased, written from a perspective that only attributes certain faculties to animals at a given time period, and so are amendable and replaceable.

To this end, I have begun preparing an essay on animal biographies that examines, for example, a giraffe's trip to France in 1827, an event that is well documented by two naturalists, one relating the animal's arrival in Marseille and the other the animal's voyage by foot to Paris. The naturalists are primarily interested in the animal's anatomy, its movements, its feeding habits, but they also write about the animal's behavior from a human perspective, which is, of course, shaped by the interests and omissions of a particular scientific practice in a particular era. Nevertheless, their curiosity is sparked by this new animal, their minds are relatively free of previous references, and their eyes are relatively open to seeing and observing. While they only write about what they wanted to say, only a part of what they saw or believed they saw, only a part of what happened, we can combine these accounts with actual ethological knowledge about giraffes, however limited this knowledge about their emotional and cognitive capacities may be. In this way, we can get closer to a part of the lived and felt experiences of the animal, while also being aware of the mutual misunderstandings between animal and men when two very different worlds meet (see figure 4).

Another example can be found in the 1947 corrida in Spain when the bull Islero conquered Manolete during a fight whose moments have been captured in a series of photographs. While these images have the disadvantage of representing only instantaneous moments and do not

*Figure 4:* Two-and-a-half-year-old giraffe, given to the French King by the Pasha of Egypt in 1827, print 1828. (Source: Bibliothèque nationale de France, used with permission)

include the smells, movements, cries, and noises of the fight, they are of great interest because they open up the possibility of analyzing these important moments, examining in close detail the animal that most spectators had hardly glanced at, preferring to focus their gaze on the torero. These images give back reality and depth to the bull; they aid in moving to the animal side. Moreover, they can be used to assess and mobilize ethological knowledge about bovines as well as historical knowledge about the evolution of corridas, the placement of blows, the importance of injuries, the postures, reactions, and resulting states; in other words, they can be used to better understand the animal's lived and felt experience as revealed in the physical.

It is appropriate to emphasize what distinguishes this scientific project from the literary attempts cited earlier. There is an important difference in the method used. These biographies, refusing a tokenistic anthropomorphism and avoiding overstatements, only relate the aspects recorded by contemporaries and add nothing to them. They offer only a slice of these animals' lives, leaving out, for example, the giraffe's life in Africa or the bull's life before and after the corrida, because these documents are too laconic and vague. Nevertheless, a link must be made with the imagination, in as controlled a way as possible, so that we come out of ourselves, our condition, so that we decenter ourselves and move to the animal side, even into the animal, in order to make ourselves (in part) animal. To try and understand and re-create, we must follow the example of field ethologists who observe and participate. At the same time, we know that despite this asymptotic approach that pushes against obstacles and moves closer to the Other, we can never fully realize the desire of bringing two worlds together. We can only bring them as close as possible, maybe even with some overlap, but our reconstitutions of animal lives remain human; they always contain some element of anthropomorphism even if this element is recognized, monitored, and decentered, just as Western ethnologists recognize their own ethnocentrism when studying non-Western human populations. Kafka understood clearly this problem when writing about the memories of an ape: "Of course what I felt then as an ape I can represent now only in human terms, and therefore I misrepresent it, but although I cannot reach back to the truth of the old ape life, there is no doubt that it lies somewhere in the direction I have indicated."[16]

Biographical work on animals should insist on the fact that for the species involved, each individual has her own journey through, and understanding of, the world, and that these journeys and understandings can change over generations, as one might illustrate using a series of canine biographies. It should also insist on the fact that each individual lives a particular encounter with humans, something that can evolve over time, as can be seen in the biographies of great apes.[17]

## AN ETHOLOGICAL HISTORY

This change over time is why the first two types of *animal* history must be integrated into a third: an ethological history[18] that places animals at the center of its approach, that is interested in the animal's own transformations, and that works from the fundamental hypothesis that species, groups, and individuals are constantly adapting to their ecological and human conditions, just as animal behaviors, sociability, and cultures are constantly fluctuating. This hypothesis is rejected by ethologists from classical and behavioral schools of thought because classical ethologists posit that behavior defines a species, and so is invariable. Nevertheless, changes in individual behavior that result in changes in group behavior within species are being recorded more and more often by field ethologists—from the discovery and spread of salting sweet potatoes in the sea by Japanese macaques to the voluntary singing of humpback whales.[19] Many ethologists no longer hesitate to speak of animal cultures within a geographical context, although they forget the temporal dimension that goes hand in hand with the geographical since individuals and then groups change their social behavior at any given moment. This approach is moreover much more in accordance with Darwinian evolutionary theory, which posits that evolution first happens through the marginal before imposing itself on the species, something classical ethology had forgotten but that cognitive ethologists are insisting on more and more as they focus on animal inventions and how these inventions spread. Even ethologists who are ready to consider and discuss the historical dimension of behaviors and cultures often do not have the necessary chronological distance because of the fact that most descriptions and usable experiments only date to the 1960s, with earlier studies being limited by the remoteness of the issues, and more importantly by the false certainty of fixed behaviors. Similarly, monographs about species, undertaken by natural history museums in the nineteenth century, do not consider individual behavior, and even less so the individual's history; they do not even consider the history of the species until the last third of the century. The historian is thus well placed for developing the idea that animal societies are not societies without a history, as was previously thought. Given that there is still very little about so-called "traditional" human societies, it should be clear that denying or attributing a history to Others is not an innocent move but rather a political one.

Herein lies the transdisciplinary interest of an ethological history that would insist on these social fluctuations, and of a historical or historicizing ethology that would allow us to study animals in a given era and then compare this to other eras, and that would consider ethology a discipline to be revisited, combined with the social and human sciences, and enriched by their concepts, as we saw earlier. Initial work in this area, recent or ongoing, by Francophone researchers demonstrates that such an approach is entirely possible. For example, with respect to wolves, Jean-Marc Moriceau has used parish registers to track the actions of anthropophagous wolves from the sixteenth to the eighteenth centuries, and reconstituted, although it was not the initial objective, a part of the ways wolves lived during that time period;[20] Corinne Beck, Éric Fabre, and Julien Alleau have created a connection between history and ecology in order to identify the areas preferred by wolves, better understand their acts and gestures, and examine the reasons for their disappearance in the nineteenth century.[21]

Such an ethological history can be combined with the work of ethologists who have undertaken a history of species from the side of animals, notably with respect to domestication, showing that this process was not imposed by humans but instead derived from an animal acceptance of a human proposition.[22] Another example from this work might be the study of the diversification of a species after domestication, such as that of dogs whose adaptations reflect different environmental and human conditions.[23] In this way, different approaches and results can be used to build a genuinely interdisciplinary history of animal behavior, its modes of construction and transmission, its temporal layers—in short, a history of animal cultures.

## NOTES


1. Éric Baratay and Jean-Luc Mayaud, "Un champ pour l'histoire: L'animal," *Cahiers d'histoire* 42, nos. 3–4 (1997): 3–4, 409–442; Éric Baratay, "Les socio-anthropo-logues et les animaux," *Sociétés* 108, no. 2 (2010): 9–18.
2. Jean-Christophe Bailly, *Le versant animal* (Paris: Bayard, 2007). Translator's note: Bailly's book has been translated as *The Animal Side* by Catherine Porter (New York: Fordham University Press, 2011).
3. Marc Bloch, *Apologie pour l'histoire ou le métier d'historien* (Paris: Armand Colin, 1997), 52.
4. Donna Haraway, *Simians, Cyborgs, and Women: The Reinvention of Nature* (New York: Routledge, 1991).
5. Yves Christen, *L'animal est-il une personne?* (Paris: Flammarion, 2009).
6. Philippe Descola, *Par-delà nature et culture* (Paris: Gallimard, 2005).
7. Dominique Lestel, *Les origines animales de la culture* (Paris: Flammarion, 2001).
8. Kevin Laland and Bennet Galef, *The Question of Animal Culture* (Cambridge, MA: Harvard University Press, 2009).
9. Florent Kohler, ed., "Sociabilités animales," *Études rurales* 189, no. 1 (2012).
10. Lucy Bates and Richard Byrnes, "Creative or Created: Using Anecdotes to Investigate Animal Cognition," *Methods* 42 (2007): 12–21.
11. Vinciane Despret, *Penser comme un rat* (Versailles: Quae, 2010).
12. Éric Baratay, *Le point de vue animal: Une autre version de l'histoire* (Paris: Seuil, 2012).
13. I adopt a larger perspective to develop this example further in *Bêtes des tranchées: Des vécus oubliés* (Paris: CNRS Editions, 2013).
14. Adam Miklosi, *Dog Behaviour, Evolution and Cognition* (Oxford: Oxford University Press, 2008).
15. Michel Pastoureau, *Les animaux célèbres* (Paris: Bonneton, 2001).
16. Franz Kafka, "A Report to an Academy," in *The Complete Stories and Parables*, ed. Nahum N. Glatzer (New York: Quality Paperback Book Club, 1988), 253.
17. Chris Herzfeld, *Wattana: Un orang-outang à Paris* (Paris: Payot, 2012).
18. Éric Baratay, "Pour une histoire éthologique et une éthologie historique," *Études rurales* 189, no. 1 (2012): 91–106.
19. Yves Christen, *L'animal est-il une personne?* (Paris: Flammarion, 2009); Naofumi Nagajawa, Masayuki Nakarichi, and Hideki Sugiura, *The Japanese Macaques* (Tokyo: Springer, 2010).
20. Jean-Marc Moriceau, *Histoire du méchant loup* (Paris: Fayard, 2007).
21. Corinne Beck and Éric Fabre, "Interroger le loup historique? Entre la biologie et l'histoire: Un dialogue interdisciplinaire," in *Repenser le sauvage grâce au retour du loup*, ed. Jean-Marc Moriceau and Philippe Madeline (Caen: Maison des Sciences Humaines, 2010), 13–21; Éric Fabre and Julien Alleau, "La disparition des loups ou essai d'écologie historique," in *L'animal sauvage entre nuisance et patrimoine*, ed. Stéphane Frioux and Émilie Pépy (Lyon: ENS, 2009), 25–34.
22. Serge Budiansky, *The Covenant of the Wild: Why Animals Choose Domestication* (New Haven: Yale University Press, 1999).
23. Raymond Coppinger and Lorna Coppinger, *Dogs: A New Understanding of Canine Origin, Behavior, and Evolution* (Chicago: University of Chicago Press, 2002).

# A Tale of Three Chameleons

## *The Animal between Science and Literature in the Age of Louis XIV*

PETER SAHLINS

The subjects of this chapter are three chameleons in Paris and Versailles represented in two different narratives early in the personal reign of Louis XIV (1661–1715). The first was the chameleon described in late 1668 by the architect, physician, and founding member of the Royal Academy of Sciences Claude Perrault (1613–1688), and featured in the *Mémoires pour servir à l'histoire naturelle des animaux* (1671). The other two were the subjects of the *Histoire de deux Chaméléons*, a work of literary naturalism written by the *salonnière* and novelist Mademoiselle de Scudéry (1607–1701) in 1673, and published in the second volume of her *Nouvelles conversations de morale dédiées au Roy* in 1688. In this paper, a microhistory of three chameleons engages an early version of French animal studies, the one announced by the anthropologist Claude Lévi-Strauss in the now-famous appraisal of the totemic phenomenon: "Animals are good to think [with]." Certainly, as Dan Sperber has remarked, not all animals, and likely not in all cultures or at all moments of history, are good to think symbolically.[1] But the chameleon, an exotic and rare animal laden with myth and metaphor, mediated the complex and ambiguous distinctions between scientific and literary discourses that took place in the salons, in the dissecting rooms, and in print culture of the French Classical Age. At the same time, the two narratives converged, unexpectedly, in thinking an opposition to a Cartesian and mechanistic model of animal identity that would deny the souls and dismiss the thought and "passions" of non-human animals, and especially of chameleons.

A Capuchin monk returning "from Egypt" in September 1668 offered the first chameleon to Louis XIV, who immediately regifted the animal to Claude Perrault and the newly founded Royal Academy of Sciences for study and dissection.[2] The chameleon lived less than a month; although its life was briefly and sparingly described, its symbolic afterlife in print and illustration was long and rich indeed. The Academy's published *Mémoires* described the bodies and anatomies of mostly exotic animals, and specifically the chameleon, in a narrative style that claimed to strip the specimen of myth, legend, and literature, while correcting the errors of previous descriptions reaching back to Democritus and Aristotle. Perrault's narratives, he insisted, established a transparent relation of the text to its referent, the chameleon's body. In the preface to the *Mémoires*, he described what was seen "with simplicity and without ornament" in an effort to "mirror" reality: "We had no other intention than to make visible things such as we saw them, as with a mirror, which puts nothing of itself, and which represents only that which has been presented (vii).[3] Similarly, Perrault downplayed the role of the artist in the visual illustration of the animal: the drawing was not a work of imagination, but an accurate representation that the Company [of Scientists] chose to represent their exact observations. This early modern principle of "objectivity," in the sense of "truth-to-nature," following Lorraine Daston and Peter Galison, informed the epistemological project of the Royal Academy.[4] Such a scientific claim was hardly new, but Perrault argued that the collectively witnessed and verified observations of the Company broke with uncorroborated observations of amateurs, of naturalists and travelers, and of classical authors, and corrected their errors in an "exact" set of observations described without metaphor or ornament.

More novel still was the Academy's rupture with a natural history of animals from a functionally anthropocentric perspective, one that studied "their habits [*moeurs*], their food, how they are hunted, their medicinal properties, and the other uses attributed to them, about which all natural historians write their books, but about which we have spoken only in passing." Rather, by focusing not on uses or behavior but on "this exact Description of its internal parts that was missing from Natural History" (v), Perrault sought to disengage the animal from its human uses, including symbolic ones, marking (perhaps) the collapse of what William Ashworth has called the "emblematic natural history of the Renaissance" with its associative framework founded on allegory and analogy. Perrault's efforts to eliminate all metaphor and ornament—all literature—in the textual representations of animals corresponds as well to Foucault's account of the "purification" of language in natural history and in the new representational episteme of the Classical Age.[5] Perrault and the Royal Academy claimed to consider the body of the chameleon independent of its symbolic identity, while avoiding metaphor, ornament, and even literature in its representations. Scholars have long noted the contradictions inherent in this project, especially its initial product: the lavish published volume of 1671 and its engravings that served as a richly symbolic monument to the glory of Louis XIV in his patronage of the sciences, and especially the new natural history.[6]

More generally, scholars have long insisted on the imbrication of science and literature in what was once called the "Scientific Revolution," studying the new forms of scientific knowledge produced in literary texts, as well as the literary dimensions of seventeenth-century scientific narratives and polemics.[7] At the same time, the advent of the French Classical Age witnessed a new discursive and institutional distinction between "science" and "literature," foremost in the professionalization, masculinization, and royal patronage of the Royal Academy of Sciences, and especially the life sciences.[8] But this sharpening boundary of the institutional spaces of science and literature, the self-conscious disciplinary shrinking (and gendering) of the scientific practice in the French Royal Academy of Sciences coincided with an expanding space of science as well: the diffusion of the new learning within the female-dominated (if not quite feminist) aristocratic and bourgeois salons in Paris. In the late 1660s, discussions in several prominent salons, along with public lectures and the beginning of public dissections, popularized in literary form the scientific discoveries of the day, helping to create what Geoffrey Sutton called a "science for a polite society."[9] Perrault's and the Academy's purely "scientific" discourse about animals, devoid of literature and symbolism, remained more notional than actual, just as the literary *salonnière* Madeleine de Scudéry's momentary engagement in the "new science," through her chameleons, was an unexpected rapprochement between science and literature in dialogue with Perrault—and against Descartes, as I shall argue.

In September 1672, the French consul in Alexandria sent Scudéry a male and female chameleon, also Egyptian, but of a species half the size of Perrault's, one of which she kept alive for seven months. In 1688, she published her *Histoire de deux Chaméléons* (the year, coincidentally, that Claude Perrault died of an infection from a dissected camel).[10] In 1672, Scudéry was a central figure in the world of literary salons in Paris. She was both celebrated and criticized as the paragon of *préciosité*, a literary and conversational style marked (in the eyes of her critics, including Boileau, Molière, and Furetière) by exaggerated refinement, misplaced prudery, superficial learning, and excessive sentimentality.[11] Her reputation has been restored by two generations of feminist scholarship that has taken seriously the philosophical work of her literary endeavors, including her ephemeral engagement in the "new science" through her chameleons.[12]

Scudéry's *Histoire* was no doubt a work of literary preciosity, a fictional conversation told in the voice of "Sappho" (Scudéry's *nom de précieuse*). The second part of the work was ever more precious: a literary cycle of poems (madrigals, epithets, and a funeral oration) composed by admirers and members of Scudéry's circle about the heroic lives and tragic deaths of the chameleons. These works, in fact, diverged little from Scudéry's own "precious histories" of a warbler (1657), doves (1664), and a pigeon (1665)—except, of course, that this was about an exotic reptile.[13] Yet the first part of her *Histoire* engaged explicitly in the "new science," and specifically in a dialogue with Claude Perrault and the "very beautiful and ample treatise that is in the hands of everyone," published the previous year. "I will not get involved in speaking about the chameleon neither as a Doctor nor as Philosopher," she wrote, excusing herself from "things she did not know." Rather, Scudéry repeatedly claimed to "simply describe what I have seen with much care and precision" (498–499). Her literary ethology, a first-person narrative describing the chameleons' diet, color changes, and behavior, tactfully corrected Perrault, much as the latter had politely dethroned Aristotle.

Claude Perrault was a regular visitor to Mademoiselle de Scudéry's salon in the early 1670s, and had seen her two chameleons alive. At their deaths, Scudéry gave Perrault their bodies for dissection, and he revised his original description, incorporating the "two small chameleons" and parts of Scudéry's narrative into a text rewritten most likely in the spring of 1673, typeset in 1688, but not published until 1733.[14] The coincidence and dialogue of the two accounts, and the textual and corporeal exchanges and appropriations of their chameleons after 1673, create an unusual laboratory to compare different specimens of scientific and literary writing about animals at the beginning of the Classical Age. In their dialogues and exchanges, the chameleons of Perrault and Scudéry illustrate the continuing hybridity of science and literature at a moment invested in their separation. At the same time, the three chameleons were also caught up in the public controversy over the reputation and writings of René Descartes, placed on the Papal Index in 1664. The king and the Church, the Paris universities and the Royal Academy of Sciences, all officially opposed and excluded Cartesian "doctrine" as a heresy linked to the Jansenist dissenters. But Descartes's followers and disciples successfully introduced a version of Cartesian philosophy to the salons and to a wider reading public. Thus Cartesian physics found wide acceptance in polite society, even if Descartes's physiology—especially Cartesian thinking about animal mechanism—did not, and the literary opposition to the "beast-machine" flourished in the 1670s.[15]

The three chameleons, in their lives and deaths, were highly visible subjects that gave life and expression to these debates between 1668 and 1673. They have not escaped the attention of historians and literary critics, although scholars have tended to insist on the distinctiveness of the two narratives by Perrault in the *Mémoires* and Scudéry in the *Histoire*, even if, following Harth and others, both had deep affinities with Cartesian "reason."[16] After examining some of the differences in the style and substance of these accounts, I will consider the unexpected similarities, convergences, and exchanges (including of the chameleons' bodies) between the two authors. As the chameleons themselves slipped between the dissecting room and the salon, they engaged a conversation with Aristotle and the Ancients and with the modern mechanistic philosophy of René Descartes: over time, the three chameleons converged in unexpected ways, and their lives and symbolic afterlives can help make sense of the reception of "Cartesianism" in France, and an unexpected rapprochement of science and literature at the beginning of the Age of Louis XIV.

## DIFFERENCES

The differences are patently obvious, and the two narratives stand radically, almost structurally opposed. Claude Perrault and Madeleine de Scudéry had different interests in their chameleons: each worked from different institutional positions of authority, in different discursive modes, with different aesthetic judgments, and with opposing ethical engagements with their chameleons, all of which informed their different and opposing understandings of the two features of the chameleon that drew their greatest attention: its diet and its color changes.

Their aesthetic opposition is notable. "In truth," wrote Claude Perrault in the original 1669 published account, "it is hard to see why the Greeks gave such a beautiful name to so vile and ugly a beast by calling it Little-Lion."[17] And Perrault could only find "altogether ridiculous" the chameleon's slow (mechanistic, subjectless) movement, "with a gravity that seemed affected because it was a gravity seemingly without a subject. That is why Tertullien said that it appeared as if the Chameleon pretends to walk rather than actually walks" (20). Perhaps it was the "automatism" of the chameleon's gait that troubled the anti-Cartesian anatomist; more likely, his negative judgment was informed by the metaphoric reputation of the chameleon that framed his description. In the unpublished scientific report to the Academy, Perrault narrated in a neutral, collective voice, without "ornament," the experiments and dissection of the chameleons, conforming to a new model of scientific narrative in the 1660s.[18] But in the published versions, despite his apologia in the preface, the description of the chameleon insisted firmly on the animal's symbolic identity. He thus began the text:

> There is hardly an animal more famous than the chameleon. Its admirable properties have been forever the subject of Natural Philosophy as well as Morality: its color changes and its reputed mode of nourishment have provided admiration across the centuries and [a subject of] inquiry for all those who have applied themselves to know Nature. And the marvels that the Doctors have told of this puny animal [*chétif animal*] have made it the most famous symbol that has been used in Morality and Rhetoric to represent the cowardly complacency of Courtiers [*Courtesans*] and Flatterers and the vanity of simpler and lighter minds [*esprits*] (4).

While the "admirable properties" and the "marvel" of the chameleon's diet of wind had made the animal a symbol of purity and virtue both in the Ancient world and the early Renaissance, the chameleon's polychromatic transformations had become—in seventeenth-century fable, morality, and literature—a symbol of deceit and flattery, and especially that of the courtier. More than a literary ornament or a particular aesthetic sensibility, the figure of the chameleon as courtier, alongside other myths and legends attached to the animal, informed the Academy's inquiry and especially its published account.

If Perrault thought the animal ugly, vile, and ridiculous (while still marveling occasionally at certain of its features), Madeleine de Scudéry was moved from the beginning with wonder and appreciation: "This animal has so many unique qualities that one cannot be amazed enough," she declared. Its gait, far from being ridiculous, was "slow, solemn, and majestic" (503). Against the wisdom of the Ancients, and Perrault's own suspicions, she refused to lower its status to that of a reptile (502), and while not attempting to class the chameleon as a bird (as some medieval bestiaries had done, elevating its moral status), nonetheless praised it as "the most beautiful animal in the world" (525).[19] The difference was more than aesthetic: it was a question of the subject. Perrault called the animal dissected by the Royal Academy "our subject," referring less to a royalist position than a scientific project of individuation and differentiation: the chameleon dissected may have been different from Aristotle's, he wrote, introducing fundamental doubt about Aristotle's claims. Scudéry turned her chameleons into true subjects—subjects with a history, a life trajectory, and a set of relationships. She thus begins with the story of their travels from Alexandria by boat to Marseille, then "in a litter to Lyon, then by carriage to Paris," where they arrived in September 1672. The exotic guests were installed, with great care, in Mademoiselle de Scudéry's hotel in the Marais, and presented to Parisian society on a regular basis. And as we shall see, Scudéry's chameleons became subjects of love, including for her.

Perrault's chameleon was never seen in public, although its portrait was painted by Pieter Boel, the Flemish animal artist employed at the Gobelins (see figure 1). The experiments on September 18, 1668, as they took place in the King's Library, were closed, even secretive undertakings by the Company of Scientists.[20] Mademoiselle de Scudéry too undertook a series of "experiments" in her salon, but these were far more public. Her salon *was* her laboratory: "All of Paris came to see" the two chameleons, including Claude Perrault. "I showed them in my room [*cabinet*] to many people, and to make them more visible, I made them climb a long silk tapestry where bands of gold on a red backdrop made them appear especially beautiful" (514). Scudéry had installed the chameleons with aesthetic and ethical care and attentiveness to her perception of their biological needs (which, of course, did not include food). She thus kept them warm at night with a "tin bottle full of hot water" but covered with a rug to prevent injury. Perrault had been less interested in keeping his alive once the initial observations and experiments were completed: as he later wrote, it died in the "first colds" of winter (on October 13, 1668), and the Company proceeded to dissect it.

The differences in aesthetics, interests, setting, and narrative between Perrault and Scudéry led naturally enough to divergent conclusions regarding the received wisdom about the chameleon, both Ancient and Modern. Perrault was far more invested than Scudéry in attempting to correct the errors of Antiquity, less Aristotle himself (who did not mention the chameleon's diet, but likely, and unusually, dissected a live one) than Pliny the Elder, Solinus, and a lost treatise of Democritus.[21] Madeleine de Scudéry, by contrast, did not cite the Ancient naturalists by name, but she did draw heavily on an account that can be traced to Pliny the Elder, repeating the ancient and enduring belief that the chameleon lived on air, and on its remarkable capacity to change colors.

For Perrault, the legend of a diet composed of wind or sunshine was easily dismissed. The Royal Academy, after all, was hardly the first to disprove substantially the myth, a fact that Perrault acknowledged only in part, citing none of the experiments or observations of Pierre

*Figure 1.* Pieter Boel, "Chameleon," ca. 1668. (Source: Art Resource; © Musée du Louvre, Dist. RMN-Grand Palais/Gérard Blot/Art Resource, NY)

Belon, Joseph Scaliger, Sir Thomas Brown, and many others.[22] But the Company's proof was to be found in a collective and ocular witness of an act "that surprised us" because of its speed.[23] The dissection of the chameleon's body was an even more incontrovertible proof. Perrault devoted much attention to the detailed anatomical description of the animal's digestive chain, from its "jaws garnished with teeth" and its remarkable tongue, to its esophagus, stomach, intestines, and kidneys—organs that were sparely described beyond their relative size and shape, but which signified a digestive regime, as displayed in the engraving by Abraham Brosse (figure 2).

Madeleine de Scudéry disagreed, politely but forcefully, with Perrault's accounts of the chameleon's diet. Despite Perrault's (and all other) evidence, Scudéry insisted that the chameleon lived only on air. Her proof? Nobody who had cared for them since their departure from Alexandria, including the ship's captain who took them out for air every day, ever saw them eat; and she herself, in seven months, did not once witness a single act of consumption. She also remained unconvinced by Perrault's description of the tongue, "itself no doubt a marvel, but these Messieurs of the Royal Academy have so well represented it that I have nothing more to add. I exhort all those who are curious to read what they have written. I will only say that in all the time that I had the chameleon, I never saw him take any flies."[24] She was persuaded that "the air and the sun's rays are their only nourishment" (although perhaps they took a

*Figure 2.* ABRAHAM BROSSE'S ENGRAVING OF THE CHAMELEON IN CLAUDE PERRAULT, *MÉMOIRES POUR SERVIR À L'HISTOIRE NATURELLE DES ANIMAUX* (1676), PLATE OPP. P. 12. (SOURCE: LINDA HALL LIBRARY)

little "viscous humor" from a grape). Even modesty did not even prevent her from describing the frequency or consistency of their excrements, which were few and far between (twice in two months for the female before she died, and four times in seven months before the male's passing), and which were "without a smell and quite firm," but hardly evidence that they did not feed on air (529).

Scudéry's fictional first-person narrative was structured by careful and exact observation: "I simply say what I have seen," a recurrent phrase that strongly recalled Perrault's own privileging of "exact observation." For Perrault, only collectively verified observation could establish objective truth, but both authors privileged the role of sight and experience. In doing so, each borrowed loosely from a "Cartesian" method concerning the pursuit of truth: to believe only those things that can be known for certain, privileging observation and experience over texts and ancient authorities. Yet Descartes (in the fourth *Méditation* [1641]) was right: sensations were "confused and obscure," and sight could be misleading. However carefully each observed their animals, Perrault and Scudéry saw completely different chameleons.

Perrault understood the chameleon's color changes within the framework of his identification of the chameleon's reputation. For the chameleon (courtier), color changes were closely linked to the sun (king): the sun changed the animal's "resting" shade of gray-blue into spots of "more brilliant colors," and when "the sun ceased to shine, the first color of gray slowly comes back and spreads across his body" (228–230). The chameleon as courtier in the Age of the Sun King dovetailed neatly with an empirical description, but Perrault was not content with a simple, mechanical relationship to the sun, either allegorical or natural. Rather, Perrault used the chameleon to respond to Descartes in hypothesizing the anatomical basis of the color changes. He dismissed the Cartesian opinion that variations could be explained as "a change in the disposition of particles which compose its skin," rejecting a purely mechanistic account of reactions to external situations. Instead he extended the humoral theory of the second-century Roman physician and philosopher Galen to animals. Perrault considered the "humors brought to the skin by the movement of the passions" as the probable physiological explanation of the color changes, which he likened to blushing in humans: "And in accordance with the skin receiving a humor capable of making it change colors, as certainly occurs when joy reddens the face of man . . . the two colors that the chameleon normally takes when he is in the sun, where he is happy, are greenish-gray and yellow."[25]

The sun plays the role not of external cause of the chameleon's color changes, but as condition of the chameleon's happiness. Apart from allegorical readings of this royalist science, it is worth insisting on the uses of the chameleon to oppose Cartesian physiology and philosophy. Descartes himself, in his metaphysical dualism of body and soul, did not in fact withhold "passions" from animals, which could in fact be, he claimed, more "violent than those of man," as these were without thought and largely corporeal, pertaining to motion, digestion, or reproduction. The "passions of the soul," by contrast, were uniquely human, as elaborated in *Le discours de la méthode* (1637) and *Les passions de l'âme* (1649), as were language, thought, and immortality.[26] Perrault's chameleon, by contrast, seemed to enjoy passions of the soul: it (he?) was not only "happy" in the sunshine, but was capable of expressing a variety of passions: "happiness," "pleasure," "joy," and "anger"—although there was no evidence of the highest passion of mid-century polite society, that of "love."

Yet Perrault's chameleon was capable of a kind of judgment or reason, as evident in the explanation of the chameleon's gait. Again criticizing Descartes, but also the received wisdom that the chameleon was a timid animal, Perrault argued that its movement could not be "pushed by fear," but was instead the "result of a great circumspection that makes him act with great caution" (21). However ridiculous the movement, however "vile and ugly" a beast, however trivial his decisions, the chameleon undertook the deliberate acts of "circumspection" and "caution." Although Perrault did not speak of the "thought" of the chameleon, it is nonetheless clear that he believed that the chameleon reasoned and lived an emotional life, however primitive in comparison with Man, the perfect animal.

Madeleine de Scudéry opposed Perrault's theory of the chameleon's color changes, but extended and deepened his understanding of animal soul and capacities. She contested Perrault's idea that the "passions" were behind the animal's polychromatic transformations, and she downplayed the role of the sun. "These changes take place without them changing location, without anything angering or pleasing them, either in the sun or in the shade." She dismissed the anthropomorphic idea that the color change was like blushing, a mechanical reaction to emotion and passion: "I have never seen evidence that this animal could have lively enough passions to believe that these color changes occur like the reddening of a man's face in anger" (527–528). But if she rejected Perrault's Galenic anthropomorphism, Madeleine de Scudéry did not hesitate to project other human characteristics onto the animal. "Passion" may not have been at the foundation of their color change, but the chameleons were highly expressive of their emotions, their desires, and their sentiments, which were all too human. She herself had witnessed "an extreme friendship between them" that even found its naive (and precious) anthropomorphic expression: "they were always holding each other with one of their little hands" (515), a fact that didn't seem to surprise her. Indeed, the chameleons were capable of love itself, what her friend and anti-Cartesian scholar of animal intelligence, Marin Cureau de la Chambre, announced in his *Discours de l'amitié et de la haine qui se trouvent entre les animaux* (Paris, 1667) as "the first of all the passions" (2).

Alas, life ended in tragedy for the female chameleon, five weeks after its arrival in Paris: during one of her demonstrations, Scudéry wrote laconically if politely, "a man of quality abruptly grabbed the small [female] chameleon and hurt her, tearing off her leg. She lived another eight or ten days; but finally she died" (514). It is clear that not all of polite society in Paris thought the same way about animals (perhaps the man in question was a Cartesian!). In fact, few did: her description of the remaining chameleon's behavior reveals a deep anthropomorphism and identification with the animal, so different from Perrault's detached judgment about the chameleon's circumspection and judgment. So afflicted with grief, Scudéry writes, the male "climbed with great haste and expediency to the top of the [cage] from where he fell three different times." Following the suicide attempt, she "redoubled her care" for the sad and afflicted survivor, and at that point named him Méléon. The proper name that abbreviated his species identification created a sentimental bond between the *salonnière* and the chameleon: "He came to love me, to know me, to hear his name and to distinguish my voice, such that I can assure those who have said the Chameleons cannot hear have made a mistake, as I have clearly seen that this one heard me, knew me, and recognized my voice, coming to me when I called him" (518). "Those who have said" included, of course, Perrault, and the observation was minimally a proof that chameleons could hear. But she also affirmed and extended, in her own precious way, Perrault's understanding of animals as sentient and even rational beings. "He sees," she declared of her surviving chameleon in a language not far from Perrault's, "with spirit and judgment, whether he walks, climbs a tree, or chooses a resting place" (507). And beyond what Perrault could ever imagine, she believed unquestionably in the chameleon's capacity to love.

Perrault and Scudéry each described chameleons of different colors—that is, two different models of the chameleon. For Perrault, it was a vile and ugly reptile whose reputation as a courtier could not be escaped, and he concluded his account, not without irony, by insisting on the continued use of the chameleon figure in the realm of "Morality and Rhetoric"—as long as its anatomy was properly understood.

> For to claim that flatterers lack candor and that vain and ambitious spirits feed on nothing, it is hardly necessary that it be true that the Chameleon takes all colors except white and that he only takes nourishment from the wind. And one can find as many subjects with which to moralize, but with more truth, based on the fact that the chameleon is without ears and almost without movement in most of his body, is only quick with his tongue from which nothing can escape, and his eyes want to see everything at once. (39)

Far from stripping animals of their moral and metaphoric identity, the anthropocentrism of the "new science" and its dismissal of Ancient myths provided a verifiable, objective foundation of a new and improved figure of the chameleon as courtier, one that befitted the Age of Louis XIV. Despite Perrault's own declarations, science swerved back to literature and metaphor, not only in the paratextual elements and illustrations of the *Mémoires pour servir à l'histoire naturelle des animaux*, but also in the published description itself.

Scudéry's text swerved as well, but in a quite different way. The narrative moved from the claim of a "careful and exact" observation of her chameleons—a literary ethology—to an expression, even a paean, to her own emotional attachments, her love for and by the "most beautiful of animals." As her story brought out the moral goodness of the surviving chameleon, in its capacity to love Scudéry, she implicitly used animals, as she had once written, as God-given "models of virtue" or a "tableau of morals."[27] She never once invoked the metaphoric identity of the chameleon as courtier, and not only because she was avoiding any negative valuation. Rather, the symbolic and human qualities of her chameleons were represented in their actual, observed behavior—their love and grief—that informed the ethical relations that Scudéry developed with her chameleons as caregiver and friend. Scudéry and Perrault engaged in different versions of anthropomorphic thinking in their literary and scientific accounts; yet despite their many differences in narrative style, method, and conclusions about the chameleon's diet and color changes, the two descriptions of the three chameleons ultimately converged.

## CONVERGENCE AND EXCHANGE

The convergence of the narratives took shape as a shared opposition of their authors to the Cartesian "beast-machine," and a shared belief in a mortal animal soul. Although they imagined the animal soul differently, they both opposed a notional "Cartesian" mechanism that reduced animal behavior to the workings of a machine, without reason, judgment, or passions of the soul, incapable of thought. They imagined the capacities of their chameleons differently, but Perrault's narrative evolved, enabled by the exchanges of the bodies of chameleons and their textual descriptions, in a notable rapprochement of science and literature.

Claude Perrault has often been described by scholars as a "Cartesian," but he explicitly declared his anti-Cartesianism in the preface to the 1680 "De la méchanique des animaux," worried that the title might be misleading. "I warn that I understand animals as beings with feelings [*sentiment*] who are capable of exercising their life functions by a principle that is called the soul; that the soul makes use of the bodily organs, which are true machines, as the principal cause of action of each one of these pieces of machinery." Perrault explained that he was interested in "the machine of the body of animals, without claims to reach higher into the research of the principle that makes them act."[28] It is true that Perrault's understanding of the functional mechanics of animal motion diverged only slightly from Descartes's, but his anti-Cartesian vitalism and animism—the belief in a life-principle of vital motion immanent or embodied in each of the operations of the anatomical machinery itself—could readily be embraced if properly explained at the anti-Cartesian salon of Madeleine de Scudéry.[29]

In the 1670s, with the publication of a half dozen treatises for and against the beast-machine that occasioned much discussion in polite society, Scudéry's salon became openly critical of the Cartesian position—although the conversation did not engage deeply or philosophically with Descartes's animal mechanism, unlike *chez* Madame de la Sablière, for example. There, "learned women" defended the writings of Pierre Gassendi and the tripartite soul of Aristotle, giving animals a "sensible" or "material" soul that La Fontaine himself elaborated in his philosophical fable against Descartes.[30] Mademoiselle de Scudéry's own literary pronouncements against Cartesianism, including her epistolary and poetic exchanges with Descartes's niece (the "fabulous Cartésie"), were less philosophical inquiries than spirited and at times precious assertions against the mechanism and automatism of animals attributed to Descartes. But her anti-Cartesian stance was not simply a reflection of her well-known love of animals. As she wrote to "Cartésie," her disagreement with the latter's uncle lay in her belief that animals had the capacity to love her:

> My belief in favor of my dog takes nothing from the esteem that I hold for M[onsieur] your uncle. For it is not my friendship for animals that makes me take note of their position, but that which they have for me that persuades me of their case. For one cannot choose to love something without having some kind of reason.[31]

As with dogs, so with chameleons. Scudéry parted company with Perrault over the belief that chameleons had a capacity to communicate with (and to love) humans. But in opposition to Cartesian automatism, and in defense of the animal soul, they shared at a minimum an understanding of the chameleon's capacity for judgment and passion.

The corporeal and textual exchanges of the chameleons (and their skeletons) furthered this convergence as their bodies moved as gifts between the salon and the dissecting room, leading to Perrault's revised text and its reworking of literature and science. In his posthumously published "Description de trois caméléons," Perrault modified his 1669 text using the "two small chameleons" that were a gift from "the illustrious Mademoiselle de Scudéry, to whom they were sent from Egypt, and who gave them to us to dissect,"[32] but also adapting and reworking some of Mademoiselle de Scudéry's textual descriptions. His engagement with Scudéry's text was polite and gracious: he included Scudéry's dissenting opinion on the eating habits of the chameleon by noting that "It was observed that the two small chameleons, during the five or six months that they lived in Paris, barely ever ate; and of all that was offered them, they only sucked several grape seeds" (46). He also softened his judgments, removing a statement in which he had described "yellow bile" as the dominant humor of the animal, and changing his description of the chameleon's gait from "ridiculous" to merely "bizarre" (37). But perhaps most strikingly, he emended Scudéry's description of the attempted suicide of her "Méléon": "When one of the little chameleons died, the other appeared to experience such a great horror, that he climbed to the top of the cage where he had been with his companion, and kept himself as far from death as he could" (37). Two elisions are notable in this account: the chameleon's love (or friendship) and the implied suicide. Everything happened as if Perrault had reached an anthropomorphizing line that he could not cross: while he might share some of the language about the reasoning capacity and the passions of his chameleons with Scudéry, he could only go so far. Yet the insertion of her rewritten text in his revision of the *Description* was itself a rapprochement of science and literature.

As for the chameleons themselves, their bodies came full circle. After dissecting those of Scudéry (at the home of the aging Valentin Conrant, secretary of the Académie française), Perrault returned the bodies to their owner, who wrote that she had them stuffed. Méléon "was filled with cotton, and as dry as he is, the color of a Dutch peddler's gray [*camelot d'Hollande gris*], mixed with other colors, can still be seen" (549). Twenty-six years later, in 1698, the English naturalist and physician Martin Lister traveled to Paris and called on Scudéry, "now in her ninety-first year, and still vigorous in mind, although her body is in ruins." "She showed me the skeletons of two chameleons that she kept alive for nearly four years; in winter she covered them with cotton, and in the coldest weather she put them under a ball of copper, full of hot water."[33] Madeleine de Scudéry's presentation of the skeletons to Lister (and account of how long they had been kept alive) may have been evidence of senility and memory loss, but her preservation of their remains was an enduring sentimental attachment and a belief in the moral, if not metaphysical, immortality of the chameleon's body.

Scudéry's literary and sentimental naturalism was born of the encounters of three chameleons between 1668 and 1673. The material and discursive exchanges of the chameleons at that moment between the Academy and the salon had helped to bring the "new science" into polite society; the exchanges also enabled a convergence of "literary" and "scientific" discourses despite different understandings of the chameleon's mental and emotional capacities—its soul. And the chameleons helped to launch the debate over Descartes's ideas of animal automatism and animal soul, the controversy over the "beast-machine," and the literary resistance of the salons to Cartesian automatism. Chameleons were indeed good to think.

## NOTES

Thanks to David Bates, Anita Guerrini, Linda Kalof, Louisa Mackenzie, Stephanie Posthumus, and especially Ramona Naddaff for their comments on earlier drafts; and thanks to the N.E.H. and the Institut d'Études Avancées de Paris for their support. Oded Rabinovitch's parallel study appeared too late to be incorporated in this chapter: "Chameleons between Science and Literature: Observation, Writing, and the Early Parisian Academy of Sciences in the Literary Field," *History of Science* 51 (2013): 33–62.

1. Claude Lévi-Strauss, *Totemism*, trans. Rodney Needham (London: Merlin Press, 1962), 101–102; Dan Sperber, "Pourquoi les animaux parfaits, les hybrides et les monstres sont-ils bons à penser symboliquement?" *L'Homme* 15, no. 2 (1985): 5–34.

2. Académie royale des sciences (Paris), Procès-verbaux des séances, vol. 4 (1668), 227r–233v; 236r–238v (the dissection); and 260r–294v (Perrault's narrative account). Perrault's literary account of the dissection was first published as the *Description anatomique d'un caméléon, d'un castor, d'un dromadaire, d'un ours, et d'une gazelle* (Paris: Frederic Leonard, 1669) in quarto edition, then reproduced among the thirteen exotic animals described in the lavishly produced folio *Mémoires pour servir à l'histoire naturelle des animaux* (Paris: Imprimerie Royale, 1671). In 1676, the royal press published a second sumptuous volume of sixteen more species: for an overview, see F. J. Cole, *A History of Comparative Anatomy: From Aristotle to the Eighteenth Century* (London, 1975), 393–442; and Anita Guerrini, "The 'Virtual Menagerie': The *Histoire des Animaux* Project," *Configurations* 14 (2006): 29–41. On the publishing history of the *Mémoires*, see Guerrini, "Perrault, Buffon, and the History of Animals," *Notes and Records of the Royal Society* 66 (2012): 393–409. The best recent biography is Antoine Picon, *Claude Perrault, 1613–1688, ou la curiosité d'un classique* (Paris: Picard, 1988).
3. The quotations in the text are taken from the unpaginated "Preface" to the *Mémoires pour servir à l'histoire naturelle des animaux* (1671).
4. Lorraine Daston and Peter Galison, *Objectivity* (New York: Zone Books, 2007), ch. 2. Roger Hahn, *The Anatomy of a Scientific Institution: The Paris Academy of Sciences, 1666–1803* (Berkeley: University of California Press, 1971), aptly names the "phenomenological positivism" of the Academicians (32).
5. William B. Ashworth Jr., "Emblematic Natural History of the Renaissance," in *Cultures of Natural History*, ed. Nicolas Jardine and Emma Spary (Cambridge: Cambridge University Press, 1996), 17–37; Michel Foucault, *The Order of Things: An Archeology of the Human Sciences* (New York: Pantheon, 1994), 31. At the same time, the natural history project reflected the "mid-seventeenth century passion for the singular and the concrete in both civil and natural history . . . an emphasis upon history over philosophy [and] observation over explanation," Lorraine Daston, "Classifications of Knowledge in the Age of Louis XIV," in *Sun King: The Ascendancy of French Culture during the Reign of Louis XIV*, ed. David Lee Rubin (Washington, DC: Folger Books, 1992), 207–220, quote 215.
6. Recent scholarship has highlighted the persistent cultural, symbolic, and literary dimensions of the *Mémoires*, particularly their illustrations by Sebastien Le Clerc: see Anita Guerrini, "The King's Animals and the King's Books: The Illustrations for the Paris Academy's *Histoire des Animaux*," *Annals of Science* 67, no. 3 (2010): 383–404; Erica Harth, "Classical Science: *Mémoires pour servir à l'histoire naturelle des animaux*," in *Actes de Baton Rouge*, ed. Selma Zebouni (Paris: Biblio 17, 1986), 208–217. My focus here is less on the illustrations and other formal and paratextual elements of the published *Mémoires* than on the narrative description itself.
7. Recent examples include Howard Marchitello, *The Machine in the Text: Science and Literature in the Age of Shakespeare and Galileo* (New York: Oxford University Press, 2011); and Frédérique Aït-Touati and Susan Emanuel, *Fictions of the Cosmos: Science and Literature in the Seventeenth Century* (Chicago: University of Chicago Press, 2011). For France, see Harcourt Brown, *Science and the Human Comedy: Natural Philosophy in French Literature from Rabelais to Maupertuis* (Toronto: University of Toronto Press, 1976).
8. Hahn, *Anatomy of a Scientific Revolution*; Erica Harth, *Cartesian Women: Versions and Subversions of Rational Discourse in the Old Regime* (Ithaca, NY: Cornell University Press, 1992); and for the later seventeenth century, Mary Terrall, "Gendered Spaces, Gendered Audiences: Inside and outside the Paris Academy of Sciences," *Configurations* 3 (1995): 207–232.
9. G. V. Sutton, *Science for a Polite Society: Gender, Culture, and the Demonstration of Enlightenment*

(Boulder: University of Colorado Press, 1995); see also Carolyn Lougee, *Le Paradis des femmes: Women, Salons, and Social Stratification in Seventeenth-Century France* (Princeton, NJ: Princeton University Press, 1976).

10. Madeleine de Scudéry, "Histoire de deux Chaméléons," in *Nouvelles conversations de morale dédiées au Roy* (Paris: La veuve de Sébastien Marbre-Cramoisy, 1688), 2:496–629; page citations in the text.
11. See for example Alain Génetiot, *Le Classicisme* (Paris: Presses Universitaires de France, 2005), ch. 2; and in more detail, Myriam Dufour-Maître, *Les Précieuses: Naissance des femmes du lettres au XVIIe siècle* (Paris: Honoré Champion, 2008), revising Georges Mongrédien, *Les précieux et les précieuses* (Paris: Mercure de France, 1963). Standard works on Scudéry and her salon include Georges Mongrédien, *Madeleine de Scudéry et son salon* (Paris: Éditions Tallandier, 1946), and more recently, Alain Niderst, *Madeleine de Scudéry, Paul Pellisson et leur monde* (Paris: Presses Universitaires de France, 1976); see also the useful study of Scudéry's animals by Nicole Aronson, "'Que diable allait-il faire dans cette galère?': Mlle de Scudéry et les animaux," in *Les trois Scudéry*, ed. Alain Niderst (Paris, 1998): 523–532.
12. Dorothy Anne Liot Backer, *Precious Women: A Feminist Phenomenon in the Age of Louis XIV* (New York: Basic Books, 1974), esp. 187–202; Joan DeJean, *Tender Geographies: Women and the Origins of the Novel in France* (New York: Columbia University Press, 1991), esp. 71–93; and Harth, *Cartesian Women*, esp. 98–106 ("A Tale of Three Chameleons").
13. Niderst, *Madeleine de Scudéry*, 374–377, 456–458, 465–469.
14. The 1688 page proofs of the "Description de trois caméléons" can be found in the Bibliothèque de l'Institut de France (Paris), Réserve Fol M 130 C***. The description of three chameleons appears in Bernard de Fontenelle, *Histoire de l'Académie royale des sciences, depuis son établissement en 1666 jusqu'à 1686*, 10 vols. (Paris: Martin, Coignard, Guérin, 1733), pt. 3, vol. 1, no. 3, and was reproduced in the 1758 (pirated?) edition printed in Amsterdam and Leipzig, to which page numbers in the text refer.
15. Stephane Van Damme, *Descartes: Essai d'histoire culturelle d'une grandeur philosophique* (Paris: Presses de Sciences Po, 2002), ch. 2; François Azouvi, *Descartes et la France: Histoire d'une passion nationale* (Paris: Fayard, 2002), 5–48; and Trevor McClaughlin, "Censorship and Defenders of the Cartesian Faith in Mid-Seventeenth-Century France," *Journal of the History of Ideas* 40, no. 4 (1979): 563–581. The best work on the literary opposition to the "beast-machine" remains Leonora Davidson Rosenfield, *From Beast-Machine to Man-Machine: Animal Soul in French Letters from Descartes to La Mettrie* (New York: Oxford University Press, 1941).
16. Harth, *Cartesian Women*, 98–106, overemphasizes the "Cartesian" affinities of Perrault and Scudéry, but usefully contrasts the official masculine, objective, and rational discourse of Perrault in the Academy with an alternative science "at the margins," by Scudéry and others, founded on a "feminine subjectivity" with a rationality that "incorporat[ed] the ethical demands of the thinking subject" (105).
17. *Description anatomique*, 4. Unless otherwise noted, all references to *Description* in the text are to the 1669 edition, n. 2 above.
18. The Academy's report was an exemplary instance of a "new narrative form": see C. Licoppe, "The Crystallization of a New Narrative Form of Experimental Reports (1660–1690)," *Science in Context* 7, no. 2 (1994): 205–244.
19. On the problem of classifying the chameleon in the seventeenth century, see Antoine Schnapper, *Le géant, la licorne, la tulipe: Collections françaises au XVIIe siècle* (Paris: Flammarion, 1988), 70–71; and more generally, Olivier Le Bihan, "Bestiaire imaginaire de l'air, du XVIe au XVIIe siècle:

Le caméléon et l'oiseau de paradis," in *Les éléments et les métamorphoses de la nature: Imaginaire et symbolique des arts dans la culture européenne du XVIe au XVIIIe siècle*, ed. Hervé Brunon, Monique Mossar, and Daniel Rabreau (Bordeaux: William Blake, Art et Arts, 2004): 139–152, esp. 141–142.

20. Académie royale des sciences, Procès-verbaux, vol. 5 (1668), 200; Elisabeth Foucart-Walter, *Pieter Boel, 1622–1674: Peintre des animaux de Louis XIV. Le fonds des études peintes des Gobelins*, Les dossiers du musée du Louvre (Paris: Réunion des musées nationaux, 2001), 91, n. 41.
21. On Aristotle's chameleon, which Jules Soury claims was the only animal Aristotle dissected alive, see "Anatomie et vivisection d'un caméléon," *Revue scientifique* no. 8 (1898); and more recently, Christopher Cosans, "Aristotle's Anatomical Philosophy of Nature," *Biology and Philosophy* 13, no. 3 (1998): 311–339.
22. Pierre Belon, *Voyage au Levant* (Paris: Gilles Corrozet, 1553), 232–233; Scaliger's account from his *Exotericatum* (first published in Paris in 1554) is cited by Edward Topsell, *The Historie of Serpents* (London, 1608), 114; Browne, *Pseudodoxia Epidemica* (London: E. Dod, 1646), ch. 21 ("That the Chameleon lives only by Aire").
23. Perrault, *Description anatomique*, 33.
24. Scudéry, "Histoire de deux Chaméléons," 520.
25. Perrault, *Description anatomique*, 40–41.
26. *Lettres de Mr. Descartes, où sont traitées plusieurs belles questions touchant la morale, la physique, la médecine, et les mathématiques* (Paris: Angot, 1657), 200; on Descartes's unresolved thinking about animal passions, see for example Sean Greenberg, "Descartes on the Passions: Function, Representation, and Motivation," *Noûs* 41, no. 4 (2007): 714–734; and C. Talon-Hugon, *Descartes ou les passions rêvées par la raison: Essai sur la théorie des passions de Descartes et de quelques-uns de ses contemporains* (Paris: Vrin, 2002).
27. Cited in Aronson, "'Que diable allait-il faire,'" 529.
28. I have used the edition by Charles and Pierre Perrault, eds., "De la méchanique des animaux," in *Oeuvres de physique et de méchanique*, 3 vols. (Amsterdam, 1727), 3:1–2.
29. Dennis Des Chene, "Mechanisms of Life in the Seventeenth Century: Borelli, Perrault, Regis," *Studies in the History and Philosophy of Biology and the Biomedical Sciences* 36, no. 2 (2005): 245–260; Paul Hoffman, "Modèle mécaniste et modèle animiste: De quelques aspects de la représentation du vivant chez Descartes, Borelli et Stahl," *Revue des sciences humaines* 59, no. 186–187 (1992): 199–211; John Wright, "The Embodied Soul in Seventeenth-Century French Medicine," *Canadian Bulletin of Medical History/Bulletin canadien d'histoire de la médecine* 8, no. 1 (1991): 21–42; and especially Francois Azouvi, "Entre Descartes et Leibnitz: L'animisme dans les essais de physique de Claude Perrault," *Recherches sur le XVIIe siècle* 5 (1982): 9–19.
30. Jean de la Fontaine's "Discours à Madame de la Sablière," likely composed around 1675, was published with the fable of "The Two Rats, the Fox, and the Egg" in his second collection of *Fables* in 1678. On the sources of La Fontaine's philosophical opposition to Descartes, see R. J. Ganim, "Scientific Verses: Subversion of Cartesian Theory and Practice in The 'Discours à Madame de La Sablière,'" in *Refiguring La Fontaine: Tercentenary Essays* (Charlottesville, VA: Rookwood Press, 1996), 101–111; and R. Jasinski, "Le Gassendisme dans le second recueil des *Fables*," in his *A travers le XVIIe siècle* (Paris, 1981), 75–120.
31. Nathalie Grande, "Une vedette des salons: Le caméléon," *Biblio* 17 (2003): 89–102, n. 98. On Scudéry's literary opposition to Descartes, see also Aronson, "'Que diable allait-il faire,'" esp. 527–529; and on the epistolary and literary exchanges of Scudéry and "Cartésie" in the 1680s, including the latter's madrigal that updated Scuderie's 1657 poem on the warbler, see the *Lettres de Mesdames de Scudéry, de Salvan de Saliez, et de Mademoiselle Descartes* (Paris: Leopard Collin, 1806), 277–295.

32. The changes were entered in the unpublished 1688 revision (see note 14); the citation (and those that follow in the text, unless otherwise noted) are from de Fontenelle, *Histoire de l'Académie royale des sciences*, 1758, 1:28.
33. Martin Lister, *An Account of Paris at the Close of the Seventeenth Century*, ed. George Henning (Shaftsbury: J. Rutter, 1823), 88–89.

# The Colonial Zoo

WALTER PUTNAM

*The fact that we lock up free and proud animals in cages is one of the most abominable practices of colonialism.*

—LOUIS ARAGON

Zoos are contact zones, areas of cultural juxtaposition and interpenetration where different animal species are put on display less to replicate any natural orderings than to confirm human conceptions of our mastery over wild nature. In that sense, the relationship between human and non-human animals remains fundamentally colonial. This paper will examine how the French colonial project intersected with the zoo that was created in Vincennes as part of the 1931 Exposition Coloniale. My premise throughout this paper will be that zoos exist primarily for human purposes, not for the benefit of animals. Bob Mullan and Garry Marvin put it succinctly: "Zoos are essentially exhibitions of human versions of the animal world."[1] They display our human understanding of animals by staging a relationship, one which places humans in a position of manifest superiority. Like all human encounters with non-humans, zoos function as representations. They are not the static, fixed type of representation we most often associate with mimetic reproduction of the real, whether in words or images. They are rather sites of kinetic, framed staging of wildlife that invite yet resist our attempts at communion and comprehension. In ways similar to our relations with other postcolonial subaltern groups, exotic animals in zoos move, perform, live, and die before our very eyes. Zoo animals undergo a double captivity, first at the hands of their captors, then through the eyes of spectators who exercise scopic power over these fetishized specimens of exotic nature. They also stand as sites of memory to the colonial past and to the postcolonial present by historicizing the displacement of countless animals from conquered territories to metropolitan capitals. They exhibit and reinforce the species boundaries that define our scientific, economic, and philosophical relationships with the natural world.

Exotic animals brought to Europe and North America served as living, visual, kinetic proof of Western hegemony over distant lands. Although they came largely during the colonial period as tokens of empire, their continued presence in the West underscores the persistence of the colonial relationship long after political decolonization had taken place. Animals, especially as captured representatives of their species, remain flagrant reminders of the epistemic violence of colonialism as it subsists in our postcolonial world. Indeed, if one takes "postcolonial" in a temporal sense, the "post" invites a misleading conclusion that we have moved beyond colonialism.

To the contrary, the institutionalization of speciesism, the death and diaspora of countless animals, the degradation and deprivation of habitat, as well as the ongoing exploitation of animals for human purposes mirror many of the worst practices of colonialism. As I shall argue, the very forces at play in the colonial dynamic live on today not only in human nations but also in the continued domination of humans over non-human animals.

The appearance of "exotic" animals was a source of amazement and wonder to many Westerners, a cause for concern and consternation for others. While the conservation of endangered species has more recently become an article of faith among zoo supporters, the original mission for zoos was twofold: to provide entertainment and to increase scientific knowledge. The capture and display of exotic species, made possible only by the deployment of extraordinary resources, served to affirm Western hegemony not just over far-flung lands and peoples but also over natural resources, animate and inanimate. Their presence posed questions of territory, property, and power in a world where, by the outbreak of World War I, some 90 percent of the inhabitable surface of the earth was either owned or controlled by imperial powers. Animals were often victims of that process, collateral damage incapable of taking a stand against human incursion into the natural world. They figure in the representational practices that led to the circulation of countless animal images from early modern times onward. Stephen Greenblatt has argued that the accelerating circulation of images beginning in early modern times created a dynamic exchange of mimetic capital that unleashed powerful forces leading to exploration and expansion: "the multiple, interconnected sites of representation, the mobility of spectacle and spectator alike, the unreality of images paradoxically linked to the dazzling power of display."[2] Animals circulated within such a system from the earliest voyages of exploration to modern times, when they unwittingly and unwillingly contributed to Western dreams of expansion and possession.

## FRAMING THE BEAST

A cover image for a book of postcards from the 1931 Exposition Coloniale in Paris placed two stylized elephants against a red background with their trunks joined to form an archway inviting spectators to visit the Pavillon de l'Afrique équatoriale (figure 1). The four living elephants brought to the colonial fair in Vincennes would form a memorable attraction for the millions of spectators who flocked to see them in a reconstructed "natural" setting. A 1948 Air France travel poster wrought in striking earth tones of brown and beige accented in wood-grain patterns showed a propeller plane flying over a herd of three African elephants. The latter seem indifferent to the arrival in Afrique occidentale and Afrique équatoriale of this airborne behemoth adorned with its trademark logo, the flying seahorse (figure 2). The images underscore the symbolic charge attributed to these iconic animals and invite a reflection on their capture, displacement, and display in Western zoos. From these graphic illustrations to the actual display of their living counterparts, we can trace an arc that has consistently reinforced human assumptions about the radical alterity attributed to non-human animals. John Berger has noted that zoos appeared at about the time when animals were disappearing from daily life in the industrialized West. He goes on: "The zoo to which people go to meet animals, to observe them, to see them, is, in fact, a monument to the impossibility of such encounters. Modern zoos are an epitaph to a relationship which was as old as man."[3] Their acclimatization

*Figure 1.* Postcard version of elephant poster for the Exposition Coloniale, 1931. (Source: private collection)

and domestication via displays in zoos made the colonial conquest tangible and legible for spectators.

Exotic animals served to define notions of identity, geography, and ideology as they staked out positions along our cultural fault lines. The lions, elephants, and giraffes captured and put on display in menageries and zoos did not, of course, come of their own volition. The domestication of wild and exotic species appealed to bourgeois values and provided visible proof of human domination over nature and its proudest specimens. Protests by the animals would have fallen on deaf ears, for as Wittgenstein pointed out, "If a lion could talk, we could not understand him."[4] These animals formed a "living museum," to use the formula from the official guide to the 1931 Exposition Coloniale.[5] In so doing, they satisfied the desire for visual witnessing that has driven epistemologies of power, subjectivity, and imagination since time immemorial. One cannot overstate the important fact that their very presence among us is the result of their forced displacement from their homeland. In that regard, zoo animals share aspects of the plight of immigrants, refugees, and slaves. Their status as property for the benefit and enjoyment of humans is underscored by the ways they are used in our most common cultural practices: food, clothing, entertainment, knowledge.

## VINCENNES

The creation of a site for the 1931 Exposition Coloniale involved carving out of the dilapidated, marginalized working-class neighborhoods of Vincennes a temporary location for the display of France's domination of its more distant outposts on the margins of empire.[6] This process of deterritorialization and reterritorialization not only characterized the colonial

*Figure 2.* Air France poster by Vincent Guerra, 1948. (Source: private collection; © Air France Museum, used with permission)

project in geopolitical terms; it also underlies to this day the fundamental relationship between humans and their others, whether human or animal. Vincennes was, in many ways, treated like a colony on the outskirts of Paris. Earlier proposals to hold the Exposition near the Jardin des Plantes or on the Champ de Mars were abandoned in favor of a more spacious, more ex-centric location. While the Jardin des Plantes was a familiar site located in the center of Paris, Vincennes in the early years of the twentieth century was remote and alien to most Parisians. Vincennes was a borderland that stood between culture and nature, between center and periphery, between civilization and savagery. It was inhabited by the poor, who lived in squalid shantytowns and embodied the fear and fascination that the Parisian bourgeoisie felt for these subjects of alterity. Its working-class population was largely seen as left-leaning and potentially seditious, best contained outside of the well-heeled center of Paris.

The site plan for the zoo created strange juxtapositions and revealing separations that were reminiscent of the organization of zoos themselves. The layout carefully segregated colonial from metropolitan zones, and the imposing monuments obstructed any view of the capital so as to create an impression for the visitor of being in another world far from nearby Paris. The colonized territories and the colonizing nations were strictly separated from each other. The central alley took the visitor "through an itinerary from the civilized splendors of Paris to the savage beasts of the zoo, with education lessons in colonial geography and ethnography along the way."[7] The subaltern status of the human inhabitants of Vincennes in their natural squalor would have created an evocative contrast to the exotic animals displayed in their well-manicured, artificial settings in the colonial zoo. Beyond the exhibits themselves, the very locale—with its imposing ramparts designed to defend Paris against the Prussians—suggested that wilderness and wildness lay on the other side of the separating boundaries. Vincennes represented the more savage, brutal pole of Paris in opposition to the refinement and sophisticated manners of Versailles. Wild animals fought at Vincennes while exotic birds paraded around Versailles.[8] Nature was left more intact at Vincennes, as evidenced in its more natural, English-style grounds and woods, which stood in opposition to Versailles's formal gardens *à la française*. The ramparts, the warren of narrow alleys, the remnants of wild nature, all suggested how Vincennes already displayed many features of the modern zoo: fearsome inhabitants living in simulated freedom but contained at a safe enough distance to be viewed by curious spectators.

## ANIMALS ON DISPLAY

The 1931 Exposition Coloniale aimed to impress its audiences through the richness of its displays, all designed to memorialize and monumentalize the French overseas empire. Alongside full-scale replicas of the temples of Angkor Wat, North African souks and minarets, and West African villages and streets, the most successful and most durable display was arguably the *parc zoologique*. Although zoos were familiar sights in cities across the world at that time, inclusion of a zoo at the Exposition Coloniale made a very explicit and telling connection between animals and the colonial project. Exotic animals were not only used for propaganda purposes, where they were put on display as living proof of the benefits of the colonizing enterprise. They became colonial subjects themselves, naturalized and nationalized as they were under the banner of *la plus grande France*. It is also important to remember that the only monuments from the Exposition to be conserved after the gates closed in November 1931 were those that contained animals: the Musée des Colonies, which housed the aquarium, and the zoo on the grounds of the Exposition itself. Both showcased exotic species from the colonies, and both attracted huge audiences desirous of viewing these specimens of the French empire. Indeed, the animals on display contributed directly to the propagandizing mission of the Exposition:[9]

> The success of the aquarium and of the pachyderms and big cats at the zoo, the only surviving witnesses of the Exposition Coloniale, illustrates the sizable challenge faced by organizers. Wanting to convince the French of the grandeur of the Empire, of the riches of its overseas possessions, of the necessity to settle and invest in these territories, they could not do better than seduce them with their animal exhibits.

From May 6 to November 15, 1931, organizers sold some 33.5 million tickets to the Exposition and another 5,288,462 tickets to the zoo alone, making it not just a popular but also a financial success. A separate two-franc ticket was required for admission to the *parc zoologique* in addition to the three-franc general admission ticket to the Exposition itself. Taking into account the variable ticketing structure and multiple entry options, Charles-Robert Ageron has estimated that approximately eight million people attended the Exposition over the course of 193 days.[10] The initial zoo population included four elephants, two giraffes, fourteen lions, one hundred baboons, nine zebras, fifteen antelopes, twelve ostriches, and over two hundred species of bird. The entire exhibit was focused on Africa and its wildlife. Situated on 3.5 hectares (8.6 acres) at the extreme southeast corner of the site, the zoo constituted the culminating point in the promenade through France's colonial territories before visitors began their circumnavigation of the Lac Daumesnil and continued on through the exhibits devoted to the colonizing powers.

The reasons for the success of the *parc zoologique* are multiple, but we can isolate three: its modernity, its innovative display techniques, and its association with colonialism. On the one hand, the Vincennes zoo seemed thoroughly modern and satisfied an appetite for novelty and scope that neither the Jardin d'Acclimatation nor the Jardin des Plantes could fill. Monsieur Bondet-Saint, writing in the *Dépêche Coloniale*, raved:[11]

> So here is revealed to the public in most impressive fashion how, in 1931, one creates a modern zoo: stunning wildlife, open air, bars replaced with moats which give visitors the impression of being in direct contact with the animals, a setting crafted for everyone. It is perfect.

By contrast, the Jardin d'Acclimatation was a venue more devoted to entertainment than science. It was best remembered for the displays of exotic humans dating from the 1870s. As we shall see, this early confluence of human and animal exhibits inaugurated the era of human zoos that lasted up to the time of the Exposition Coloniale. The Jardin was also notorious for the grisly episode during the Prussian siege of Paris in 1870–71 when many of the animals were slaughtered and served to feed hungry Parisians.

The menagerie at the Jardin des Plantes was a historical and cosmopolitan landmark, but it was hemmed in by the larger botanical garden, the Muséum National d'Histoire Naturelle, and then by the city of Paris itself. The inability to expand as well as the use of more conventional ways of displaying animals in barred cages made it appear outmoded and even carceral in comparison to its more modern rival. Paul Escudié, a *député* for Paris, wrote a scathing article in 1911 titled "L'enfer des bêtes" ("Animals in Hell"), in which he denounced with great vitriol the lack of care given to the noble exotic animals rotting in their humid, filthy cells at the Jardin des Plantes. Gustave Loisel, distinguished professor of zoology and historian of menageries, noted in less acerbic terms that the Jardin des Plantes suffered from outdated buildings, insufficient trained personnel, and overpopulation of animals for the limited space available.[12] The menagerie at the Jardin des Plantes also had a scientific vocation as a unit of the Museum of Natural History; while it was a popular destination for Parisians, the Jardin des Plantes was still very much run by zoologists and naturalists.

The directors of the menagerie were snubbed when the contract for the zoo at the Exposition Coloniale was awarded to the Hagenbeck firm. The choice of Hagenbeck over the scientists of the Jardin des Plantes was a clear endorsement of attention-grabbing layouts and viewer-friendly display techniques over scientific technocracy. They did, however, participate actively in the creation of the permanent Zoo de Vincennes that emerged two years later from the temporary zoo at the Exposition Coloniale. In 1934, the Jardin sold or transferred 192 animals to the new zoo, rare specimens joined by Hagenbeck's stock of animals from the temporary zoo that were purchased for the new facility. The first director of the Zoo de Vincennes was none other than Achille Urbain, second in command at the Jardin des Plantes and chair of ethology at the Muséum. His appointment served to bring the rival institutions under a common leadership. When the Vincennes zoo was inaugurated on June 2, 1934, in the presence of the French president, it brought in over seven million francs in less than seven months.[13]

Professor Urbain would himself conduct two important missions, one to North and West Africa, the other to Indochina, in order to select animals for the new museum's collections.[14] From this latter voyage, he brought back over 250 mammals and 125 rare birds that left Saigon in 1937 on a cargo ship bound for their new domicile in Vincennes. His ark carried elephants, panthers, buffalo, macaques, gibbons, pheasants, storks, and countless other specimens.

The zoo supported and validated the colonial project through its presence at the Exposition. This propagandizing role was all the more important since, as Raoul Girardet has demonstrated, large segments of the French population were skeptical about the overall benefits of the overseas empire. The 1931 Exposition Coloniale arrived at the apogee of France's colonial consciousness, where the idea of *la plus grande France* of a hundred million citizens appealed to nationalist sentiments and universalist ambitions. Paul Reynaud's inaugural address captured this spirit:[15]

> The main goal of the Exposition is to make the French people aware of their Empire, in the words of the Convention. Each of us should feel like a citizen of this greater France that stretches to the four corners of the world . . . Metropolitan France has the second largest land mass in Europe after Russia yet she is only one twenty-third of the size of the French empire.

Despite the triumphalism and jingoism of official rhetoric, there were opposing voices from socialists, surrealists, and internationalists. Opponents of French imperialism organized a counter-exhibit in September 1931 where they drew on witnesses and accounts of colonial abuses such as those of Albert Londres and André Gide. The surrealists famously brandished the slogan "Do not visit the Exposition Coloniale." Although these protests did not directly address zoo animals, Louis Aragon intuitively hit upon a striking parallel: "The fact that we lock up free and proud animals in cages is one of the most abominable practices of colonialism."

The Exposition and zoo at Vincennes staged colonial conquest in obvious but also unusual ways. Homi Bhabha has observed: "Colonial power produces the colonized as a fixed reality which is at once an 'other' and yet entirely knowable and visible."[16] The colonial zoo made animals into visible signifiers of radical alterity in the eyes of spectators who could relish in the power and reach of their empire. Since Roman times, the display of exotic animals has become synonymous with victory over the enemy. The exotic creatures exhibited in zoos of the nineteenth and twentieth centuries seemed to appear by magic, conjured up from nowhere, or at least nowhere that most Europeans had ever visited. Their sudden, improbable eruption within European signifying systems caused a mélange of wonder and desire. They created a visual, corporeal connection to a mysterious, mythical place of origin, and by association and extrapolation came to represent that fabulous homeland in the popular imagination. These animals loaned themselves to imaginary origins and fabulous genealogies. Why else would the giraffe have been imagined to be a hybridized cross between a camel and a leopard, reflected in its scientific name, Camelopardalis? Although lions and elephants were not unfamiliar to Europeans through their long-standing association with royalty, many species such as giraffes, kangaroos, and exotic birds were objects of fascination and wonder to more recent viewers. Rather than inspiring fear, they served to incite desire—more specifically, the desire to reestablish the connections between origin and destination, between there and here, between past and present. The arrival of legions of exotic animals on European shores is closely connected to the inverse journey of Europeans setting out to discover and conquer the vast world from which they came.

## HAGENBECK'S PARADISE

Carl Hagenbeck had built a veritable empire as the world's premier purveyor of exotic animals to the circuses and zoos that were proliferating across Europe and North America from the 1870s onward. Indeed, a full-page advertisement for the Hagenbeck firm sits opposite the article devoted to the zoo in the official guide to the Exposition, offering any sort of animal for zoos, circuses, and hunts.[17] Thousands of animals—lions, tigers, elephants, rhinoceros, and every imaginable species of bird and reptile—entered Europe through Hagenbeck's vast network after being hunted, captured, or bought from locals in Africa and Asia.[18] The animal trade, for all the wonder and information that it provided, represented a veritable Middle

***Figure 3.*** Crowd in front of the elephants in the Vincennes zoo. (Source: © Albert Harlingue/Roger-Viollet/The Image Works, used with permission)

Passage for the animals, with low survival rates and miserable conditions along the way to a life of captivity. Estimated losses range between 50 and 90 percent.[19] Comparable in many ways to the human slave trade, the forced removal of countless animals, considered as property by their owners and transporters, has a long history that precedes the Exposition Coloniale.[20]

The scope of Hagenbeck's impact on the modern zoo is hard to overstate. Whether it is the commercialization of the animal trade or the innovative "natural" zoo design without visible bars and enclosures, Hagenbeck shaped the way we view animals in myriad ways. The *parc zoologique* in Paris, like many others, was based on Hagenbeck's Tierpark, opened in 1907 in Stellingen, near Hamburg. Rather than confining animals to cages, he built a series of platforms and natural boundaries, especially moats, that created the illusion for spectators that the animals were free to roam about their natural landscapes. "Hagenbeck called his creation an 'animal paradise,' associating in his mind—and for the public—his park and the biblical Paradise, a place where 'animals would live beside each other in harmony and where the fight for survival would be eliminated.'"[21] But Hagenbeck's biblical Paradise constitutes a misnomer for the captive animals who must endure the gazes and taunts of humans as reminders of their powerlessness. John Berger has written eloquently about the marginalization of zoo animals and the impossibility of a reciprocal gaze across bars and moats.[22] Hagenbeck's innovations sought to display captive animals in a way that makes the technology of captivity invisible. This rhetoric would be echoed in the official guide to the Exposition, where visitors were reassured that only hungry lions would attack defenseless creatures, but that free and unprovoked lions would live peacefully alongside goats and antelopes with little danger for the latter.[23] Hagenbeck himself had nevertheless devoted considerable time to measuring the distance and height that each animal could jump in order to build adequate grottos and moats, with or without water, in order to contain any wild beasts who had not read the official guide.

Hagenbeck considered these animal exhibits as *tableaux vivants* where living, moving creatures were placed in carefully constructed arrangements, always with the viewer's perspective in mind. The still-life corollary to Hagenbeck's animated compositions would be the dioramas that proliferated at the same time in natural-history museums around the world; of course, much of the interest generated by zoos was that they were "veritable living museums of exotic creatures."[24] In the classic iteration of the *tableau vivant*, the panoramic and hierarchical view is privileged. In Hagenbeck's zooscapes, this effect was obtained by placing a large rock formation in the background with animals arranged in hierarchical order, generally with the more docile species up front and the more aggressive animals in the background.[25] The famous Grand Rocher that towers over the Vincennes zoo stands not only as a symbol of the zoo but also as a background focal point around which the animal exhibits are organized. From ancient times through Renaissance masters and up to European and American landscape artists, the use of a large natural element such as the Grand Rocher has figured in the background of countless paintings as a framing and perspectival anchor. It is the backdrop against which the viewer's line of sight can be established, meant to keep the viewer's eye on a horizontal plane with lines leading to a distant focal point. From ground level, the viewer's perspective is built around the ability to collapse discretely separated species into a seamless whole where different animals appear to cohabitate and to interact. The zooscape strives to suggest a totalizing vision of nature in the eye of the spectator. The rock formations, made of reinforced concrete, as well as the vegetation, moats, and fencing are designed to be as realistic as possible in color and texture. As John Berger has pointed out, the convention of perspective used in European art since the Renaissance places the beholder in a unique position: "Perspective makes the single eye the centre of the visible world. Everything converges on to the eye as to the vanishing point of infinity. The visible world is arranged for the spectator as the universe was once thought to be arranged for God."[26] In similar fashion, Hagenbeck arranged animals in evocative perspectives that placed the eye of the spectator in a God-like position of power over all the creatures within his visual grasp. Although they moved, the animals of Hagenbeck's zoo were limited in their range and were admired for their ability to create living tableaux of nature that shifted before their spectator's eyes. The operation resembled a theater or ballet stage with the animal performers playing the roles of actors or dancers moving across a contained space for the enjoyment of the audience. One could thus produce an impression of looking out over an African landscape with birds, quadrupeds, and carnivores all sharing the same environment yet not competing for the same territory.

## AT PLAY IN THE FIELDS OF LANGUAGE

Exotic animals become intertwined in colonial discourse both as living creatures and as cultural and linguistic signifiers. Metonymy is the preferred trope of colonialism as it is of animals, and their convergence in the zoo provides an opportunity to examine them jointly. In their simplest expressions, metaphor works by substitution and metonymy works by contiguity. Structural linguists, following the insight of Saussure, distinguish between two axes, paradigmatic and syntagmatic, corresponding respectively to operations of substitution and sequential ordering. In this scheme, metaphor functions by selection, metonymy by combination. Animal metaphors—"he is a snake," "she is a fox"—most often rely on widely recognized characteristics of an animal imputed to a human, whether derogatory or laudatory. Metonymy works not by a direct substitution but rather by a suggested common association derived from close proximity or contiguity. Metaphors must create distance between two terms in order to make the substitution feasible; metonymies must maintain both terms present and proximate in order for them to create a dynamic, reciprocal meaning. In the signifying field of animals, this means that metaphors supplant one term for another. Jonathan Swift seized on the symbolic potential of animals in Western representations of the colonial world:[27]

> So Geographers in *Afric*-maps
> With Savage-Pictures fill their Gaps;
> And o'er uninhabitable Downs
> Place Elephants for want of Towns.

It is no coincidence that countless maps of blank continents were filled with animal signifiers until such a time as Western explorers and colonizers could make them known.[28] The "discovery" of exotic animals and their subsequent displacement to the zoos of Europe and North America can be characterized as a shift from metaphor to metonymy. The wild animals on display in zoos came to represent their places of origin by the many associations that were imputed to them. They were living fragments of faraway lands that became overdetermined signifiers of their colonial places of origin: fierce, enigmatic, colorful, strange, seductive, etc. Spectators could impute many qualities to them, and infer myriad meanings back to their place of origin. These fragments of empire soon came to stand in for the whole, representing more than they could possibly contain to viewers who could only imagine the contexts from which they originated.

## HUMAN ZOOS

The colonial exhibition of wild animals was often staged alongside displays of exotic peoples in what became known as "human zoos."[29] Their conception and execution bear many resemblances to animal displays and speak volumes to the status of both human and non-human animals in later nineteenth- and early twentieth-century Euro-American culture. The proliferation of these exhibits coincided with the outward expansion of European imperialism, especially in Africa following the 1885 Congress of Berlin. The scramble for Africa also had an impact on animals, both as collateral damage in the pursuit of natural resources, and as tokens of empire to be relocated to the metropolitan centers and put on display in zoos and circuses. Wild-animal spectacles soon became commonplace, but zoo attendance ultimately waned as audiences came to expect more novel, more unusual exhibits. Beginning in 1874, Carl Hagenbeck introduced "primitive peoples" alongside his animal exhibits; spectators returned in droves to his Hamburg zoo to see Nubians, Samoans, and Inuits presented as authentic specimens of wild human nature. Their presence confirmed and challenged the evolutionary and racist discourses that were so prevalent in the latter decades of the nineteenth century. The juxtaposition of human and non-human animals was meant to suggest that these primitive peoples were closer to animals according to some hypothetical evolutionary scheme. It also unwittingly cast animals as species frozen in time, hierarchically inferior, and disposable.

The categorization and commodification of animals is based on an epistemology of racial difference that mutates into a hierarchy of species difference. Speciesism, like racism, supposes a fundamental difference of kind, not degree. The exhibition of "savages" alongside "wild animals" insidiously suggested that there were inviolable characteristics of both groups that made them inferior. The popular success of these human zoos soon spread to France. In 1877, Geoffrey de Saint-Hilaire presented his own ethnographic spectacle of Nubians and Inuits at the Jardin d'Acclimatation in Paris. Attendance doubled over the following months, and a popular phenomenon was launched, one that would culminate in the 1931 Exposition Coloniale. These ethnographic shows followed many of the precepts of zoos, reinforcing the widespread assumption that these human specimens were not so removed from their animal companions. They were staged in alleged natural settings, and viewers were invited to suspend disbelief and accept their behavior and activities as normal and representative. "Scène typique" became the descriptor of choice to underscore the idea that viewers were contemplating one of countless identical native villages on display. Primitive huts were constructed, sometimes from materials brought with the natives from their homelands; daily activities such as cooking and washing were carried out on a regular schedule. Dances and rituals were performed with all the allure of authentic, spontaneous events. As was the case for animals, these natives were purported to be doing what was natural and normal to them, as if on stage without acknowledging that throngs of spectators were gawking at them. The burgeoning colonial fairs that proliferated across Europe and North America for some sixty years resulted in thousands of Senegalese villagers, Kanak cannibals, and Osage Indians being put on display. These fairs were in decline by the 1931 Exposition Coloniale when Europe was turning its attention to the specter of fascism and when other more technological means of capture and display were rising to the fore.

## KING KONG

The closing chapter in the human zoo phenomenon coincided with the 1931 Exposition Coloniale, which itself coincided with the widespread fascination with moving pictures. Just as Hagenbeck and Saint-Hilaire had introduced human exhibits in response to declining attendance at animal zoos, the public soon turned its attention elsewhere. The thirst for ever greater and more powerful images saw the appearance on the international scene of an even more formidable rival: *King Kong*. The classic retelling of the Beauty and the Beast story premiered in 1933 and demonstrated the powerful potential of the mediated capture and display of animals. Although live animal exhibits would continue to attract spectators, the allure of cinematic representation presented opportunities and challenges to traditional ways of looking at animals. Film and animals make an ideal pair given their reliance on motion and the fact that a scopic public seems irresistibly drawn to them. The Lumière Brothers, like Thomas A. Edison, had made some of their earliest films at ethnographic and colonial exhibits. Starewicz and Muybridge had used film technology both to create stories using animals and to provide scientific studies of animal locomotion. *King Kong* stages the capture and display of a larger-than-life creature, part human, part animal, who rebels against human constraints and limitations with tragic consequences. Created by humans, he reacts like a human might under similar circumstances.

Jonathan Burt observes that "film's emphasis on action and event was from a spectator's point of view much closer to the ideal zoo exhibit and provided contrast to those many hours when actual zoo animals do very little and zoo exhibits are minimally eventful."[30] Cinema (from the Greek *kinéma*, meaning movement) allowed the capture and display of the kinetic quality that fundamentally characterizes animals. The staging of animal movement so avidly sought by circuses and zoos became possible thanks to the medium of film. Whether filming animated or realistic representations of animals, cinema manages to add another layer of distance between spectator and spectacle. It freezes the spectator in a fixed gaze at the screen. It practices its own form of the colonial relationship, which, like the human-animal relationship, is characterized by a tension between proximity and distance. The need to see the collected and constructed objects of colonial display lives in a paradoxical and ironic space between spectator and spectacle. This scopic distance highlights the need for colonial artifacts to be visually affirmed in order better to underscore the hierarchical superiority of the viewer over his domain. They would, of course, remain objects in the minds of most spectators, even when filmmakers attempted to attribute to them some modicum of subjectivity by having them talk or express emotions in human terms.

Anthropomorphism became even more pervasive thanks to the movie industry. Animals could also join the ranks of narration since writers and editors could insert them into structured, constructed stories. The cinematographic image provided access to native locations and habitats, even when artificially constructed on film sets. In a sense, the pictorial ambitions of Hagenbeck described earlier find their next stage of development on the silver screen. Animals have been a staple of the movie industry since the days of *King Kong*, either in bit parts or as main actors. One can hypothesize that the more recent success of reality shows has capitalized on the scopophilic, even pornographic, visionary allure that made zoos, human or animal, such popular attractions in modern urban spaces.

## RENOVATED ZOO

The Vincennes zoo had welcomed some 89 million visitors when it closed down in 2011 for a sweeping renovation.[31] The novelty of Hagenbeck's "natural" landscapes had declined, the infrastructure required major work, and attendance had been waning.[32] The decision was made to transform the site entirely, except for Le Grand Rocher. When it reopens in 2014, the new zoo will contain five biozones on the 14.5 hectares of the original zoo: Europe, Patagonia, the Sahel, Madagascar, and Guyana. A future zone for sub-Saharan Africa is planned, as is an off-site area devoted to Australia. The overarching themes of biodiversity and conservation will be highlighted in each zone, where visitors walk through constructions of natural environments and habitats. The jewel of the European exhibit, next to the Grand Rocher, is a 4,000 m$^2$ glass enclosure for tropical birds, flora, and fauna. While each biozone will display animals from their respective geographical areas, the previous zoo's goal of having every imaginable species represented has given way to a more selective, more clustered array of species, most of which are exhibited together in their biozones rather than separated into taxonomic categories. The avowed organizing principle of the renovated zoo is to place the animal in the center of every decision, re-creating simulated experiences of searching for food and moving about that replicate as nearly as possible their original, natural behaviors. Many of these animals, of course, have never lived in their original habitats, since they were bred in zoos and traded or purchased from one captive setting to another.

The new zoo projects about 180 species and over 1,000 animals, but the setting will be as important as the animals themselves. There will be 40 percent more vegetation in the new zoo, where enclosures and separations will be minimal or invisible. "Immersion" and "integration" are key words in describing the experience of the new Zoo de Vincennes. Visitors are expected to immerse themselves in the animal's habitat and observe it living in a natural environment, as opposed to the more traditional unidirectional gaze that characterized the former colonial zoo. The emphasis on biozones is a reminder that the earlier zoos had a dual mission of knowledge and entertainment to which contemporary zoos have added the imperatives of conservation and protection of wildlife and habitat. One can safely predict that this new chapter in the Zoo de Vincennes will meet many challenges that will mirror the changing relationship between human and non-human animals.

## NOTES

1. Bob Mullan and Garry Marvin, *Zoo Culture* (London: Weidenfeld & Nicolson, 1987), 68.
2. Stephen Greenblatt, *Marvelous Possessions: The Wonder of the New World* (Chicago: University of Chicago Press, 1991), 6–7.
3. John Berger, *About Looking* (New York: Pantheon, 1980), 21.
4. Ludwig Wittgenstein, *Philosophical Investigations* (Oxford: Blackwell, 1953), 213.
5. A. Demaison, *Guide officiel à l'Exposition Coloniale Internationale* (Paris: Éditions Mayeux, 1931), 123.
6. This section is indebted to chapter 4 of Patricia A. Morton's *Hybrid Modernities: Architecture and Representation at the 1931 Colonial Exposition* (Cambridge, MA: MIT Press, 2000).
7. Morton, *Hybrid Modernities*, 17.
8. For a thorough account of the Versailles menagerie in opposition to Vincennes, see Peter Sahlins, "The Royal Ménageries of Louis XIV and the Civilizing Process Revisited," *French Historical Studies* 35 (2012): 237–267.
9. Catherine Hodeir and Michel Pierre, *L'Exposition Coloniale* (Paris: Éditions Complexe, 1991), 88.
10. Charles-Robert Ageron, "L'Exposition Coloniale de 1931: Mythe républicain ou mythe impérial?" in *Les lieux de mémoire*, ed. Pierre Nora, vol. 1, *La République* (Paris: Gallimard, 1984), 577. Despite the global economic depression and especially bad weather, the Exposition managed to show a profit of 30–35 million francs and the zoo brought in some 10 million francs by itself, a tidy return on a 5 million franc investment (Hodeir and Pierre, *L'Exposition*, 87).
11. Quoted in Pierre Loevenbruck, *Animaux captifs: La vie des zoos* (Paris: La Toison d'or, 1954), 89.
12. Yves Laissus and Jean-Jacques Petter, *Les animaux du Muséum, 1793–1993* (Paris: Muséum National d'Histoire Naturelle, 1993), 173.
13. Laissus and Petter, *Les animaux*, 186.
14. Loevenbrock, *Animaux captifs*, 95.
15. Raoul Girardet, *L'idée coloniale en France de 1871 à 1962* (Paris: La Table Ronde, 1972), 185.

16. Homi K. Bhabha, "The Other Question: Difference, Discrimination, and the Discourse of Colonialism," in *Black British Cultural Studies: A Reader*, ed. Houston A. Baker Jr., Manthia Diawara, and Ruth H. Lindeborg (Chicago: University of Chicago Press, 1996), 93.
17. Demaison, *Guide officiel*, 122.
18. For an account of the staggering number of animals traded by Hagenbeck in his first twenty years, see Nigel Rothfels, *Savages and Beasts: The Birth of the Modern Zoo* (Baltimore: Johns Hopkins University Press, 2002), 58.
19. Éric Baratay and Élisabeth Hardouin-Fugier, *Zoo: A History of Zoological Gardens in the West* (London: Reaktion, 2002), 118.
20. For a full discussion of the analogy between animal treatment and the slave trade, see Marjorie Spiegel, *The Dreaded Comparison: Human and Animal Slavery* (Philadelphia: New Society, 1988).
21. Rothfels, *Savages and Beasts*, 163. The term "paradise" descends from an old Persian root that originally designated a walled enclosure and formed the basis for the Greek word *paradeisos*, which referred explicitly to a walled-in space for containing animals.
22. Berger, *About Looking*, 21–28.
23. Demaison, *Guide officiel*, 124.
24. Demaison, *Guide officiel*, 122.
25. For an illustration of Hagenbeck's Zoological Paradise, see Rothfels, *Savages and Beasts*, 166.
26. Berger, *About Looking*, 16.
27. *On Poetry: A Rhapsody*, 1733.
28. For an illustrated study of this phenomenon, see Wilma George, *Animals and Maps* (London: Secker and Warburg, 1969).
29. For a longer discussion of this topic, see Walter Putnam, "Please Don't Feed the Natives: Human Zoos, Colonial Desire, and Bodies on Display," *French Literature Series* 39 (2013): 55–68.
30. Jonathan Burt, *Animals in Film* (London: Reaktion Books, 2002), 19–20.
31. For a complete explanation of the principles behind the Zoo de Vincennes renovation, see their description online at http://parczoologiquedeparis.fr/le-projet/un-zoo-pour-la-biodiversite/un-endroit-pour-montrer-et-conserver.
32. Visitors declined from 1.5 million in 1968 to 1 million in 1984; 600,000 in 2004; and only 300,000 in 2005.

PART 2

# Animal Philosophies and Representations

# The Unexpected Resemblance between Dualism and Continuism, or How to Break a Philosophical Stalemate

FLORENCE BURGAT

*Translated by* DANTZEL CENATIEMPO

## WHAT IS ANIMAL PHILOSOPHY?

### Reasons behind the Wording

Let us focus first on the term itself: "animal philosophy." Although the syntax of this expression, whose meaning we aim to better understand, may be disputable,[1] it is useful despite its shortcomings, especially when compared to the more exact terminology "philosophy of the animal." In this latter expression, the use of the singular "animal" to designate a multitude of animal species that often have nothing in common has become contentious. Hence Jacques Derrida's critique of the singular form—"The Animal, as they say"[2]—the official designation passed down from age to age that allowed human subjects to reduce and undervalue an infinite diversity so as to more fully serve their own interests. Any future, naive use of the term is no longer valid.

Under the pretext of a mere descriptive designation, the singular "animal" reduces a multiplicity of ways of being, saying, and doing to one single unity, further demoting the animal (and here all pretense of harmlessness disappears). In other words, one speaks of the animal in order to better forget all animals. Some will perhaps argue that the singular form, in its capacity to express the universal, the essence, is the only term that can convey the power of such a concept, while the plural remains handicapped by too great a variety of potential meanings. However, in this case, the plural designates the vitality of a thousand irreducibly different life forces—the vitality of birds who streak the sky with their calls, of dogs who wait patiently, of bees who zigzag along in the sunlight of early spring, of toads who remain motionless as relics from the beginning of time. It should seem fairly clear that the singular term "animal" would not have been the target of Derridean criticism if its philosophy had been truly cognizant of the singularity of each animal behind the classification of all animals in general. At the heart of such language about animals (here we refer to philosophers but could also mention scientists who reify the object of study in order to meet the reductionist requirements of their field), there is a key structure to dismantle, that of the establishment of a center (the subject) and its periphery. And yet we must not get ahead of ourselves, since it is far from certain that everything filed under the heading of "animal philosophy" adopts the same approach, or even the same objective of deconstructing humanism's idea of what is unique or proper to the human, and critiquing the supposed eminent dignity that follows from it.

As for the "philosophy of animality," this expression threatens to obscure the very issue it seeks to reveal, by its patent use of the term "animality" to designate the limited representation that humans have of animals. Doesn't animality promote the Bataillean notion of a basic disorder or dark reservoir of uncontrollable impulses, sexual deviance, and "bestial" violence that one must "accept" from a post-Darwinian and psychoanalytic perspective? Must humans "accept their animality"? To refer to the "animal side of man" or his "animality" is today viewed as an avant-garde position, a shocking discovery, a truth that must be admitted no matter how hard the blow to human pride. And yet, how can this proposition constitute anything less than a new way of demeaning animals, as well as a reincarnation of the classic dualistic conception of man as rational animal—only this time insisting on the animal side of his nature rather than on the rational side? What, then, could constitute this "animal side of man"?

One might object to this concept of animality—and quite differently than would an anti-Darwinian who supports the theory of human creation as separate from animal evolution, and who upholds the philosophy of a metaphysical humanism—because the supposedly continuist viewpoint eventually reaches the same conclusion it purports to oppose. In essence, it advocates a sort of primal layer of consciousness, something deeply archaic and universal that all animals naturally have, and of which humans are also now finally bearers. But is this not just the physical body that yields to an "animal logic" while the human works to overcome such urges? Some psychoanalysts have affixed the label of "animality" to our subconscious mind, as if doing so will shed light on it. Here again rises intact the definition of man as rational animal, but from the opposite side! Instead of observing ourselves from the summit of reason, we more modestly pretend to ponder the shadowy side of our nature. It would be difficult to find a more thorough way to emphasize yet again the singularity and superiority of human beings, and to offer at the same time a more dull and uninspiring definition of the animal shamefully hidden away inside us. And yet we believe that we pay tribute to this side of human nature by using the term "animality" to allude to those darker impulses over which we so often lose control. Strange tribute indeed.

## An Emerging Field

The expression "animal philosophy" has begun to be used to designate a field of research that is both new and independent. It seems fair to say that the creation of a category for "animal philosophy" is due in large part to a body of recent publications that explore "the animal question"—a term that has also made its way into official philosophical language. One can date the birth of this new philosophical work on animals, the animal, and animality in France. Published in 1978 in a remarkable and innovative edition of the journal *Critique* that was entirely devoted to "Animalité," Élisabeth de Fontenay's article "La bête est sans raison" opened up an unexplored field and can be said to represent the beginning of a cohesive discipline.[3] The author of the now classic *Le silence des bêtes*, which truly puts Western philosophy as a whole to the test, brought this beginning to fruition.[4] What happens to philosophy when it attempts to discuss animals, how does it begin, where can it go? Has philosophy even been able to ask itself these crucial questions? This is not to say that the concept of animality or of the animal has been entirely absent from the history of Western philosophy, nor that the moral issues related to the treatment of animals have been completely ignored—one need only recall Plutarch's many treatises on animals, in particular his *Whether It Be Lawful to Eat Flesh*, or the quarrel concerning animals' souls and the theological challenge posed by the concept of animal suffering, which, when it emerged, did so in opposition to Cartesianism. But the essential question, with all its attendant presuppositions (ontological, epistemological) and its implications (moral, legal, practical), deliberated as a question that encompasses all others because of how deep it goes into the foundation of everything we construct from our thought processes to our practical life—this question is new. It is an issue that topples our most fundamental notions about action, speech, and thought as no other issue could. It is a question so profound that only philosophy is equal to the challenge of investigating it, especially when compared to the relatively less demanding topics and ideas currently popular in the social sciences and humanities. Literature, of course, is not included in this reproach, and in fact it is to literature that one can turn to find the most powerfully relevant questions and experiences. Yet, philosophy's superiority in this domain can be defended by noting that the humanities and social sciences, if we may judge by the work of certain authors, are attempting to limit the "animal question" to an anthropological relationship between man and animal, usually for the purpose of preserving their traditional roles.[5]

In roughly twenty years, philosophical work as it concerns animals (to which our discussion is limited for purposes of this topic) has been developed, deepened, ramified, and radicalized as well. After the period of asking questions about anthropozoological differences—a vast area that invites detailed examination, as well as an essential topic for building new theoretical foundations—came the period of examining how these differences were used as a mechanism for excluding animals. The ontological poverty of animals that such differences posited led naturally to their entire exclusion from the moral sphere, and consequently the legal-political sphere. It became clear that much work needed to be done, with the first task at hand being to undo, to take down, to dismantle these modes of thought.

We cannot undertake a full bibliographical review of the philosophical work that has been accomplished since then; suffice it to say that the entirety of the subject matter and its major issues (ontological, moral, epistemological, etc.) have been explored. The diffusion of Anglo-American philosophy throughout France, thanks to the French translations of some major works,[6] as well as additional work fostered by these lines of thought,[7] has forced French philosophers to take into account current prominent theories, usually with the ultimate goal of refuting them. Unfortunately, some contemporary scholars have simply presented reductive summaries of philosophies that they haven't even read firsthand, as if such knowledge was sufficient to sustain serious philosophical discussion. The debate around the Great Ape Project,[8] which raged in the appropriately named journal *Le Débat*, perfectly reflects a certain disdain for American philosophy among French philosophers. By basing their argument on only a few elements rather than on a sound knowledge of the philosophy to which they were responding, some of these French philosophers misrepresented Paola Cavalieri's position by referencing only one of her articles, the one whose French translation was found in the journal.

At the same time, it is clear that giving a name to the research field of animal philosophy, which was until recently the domain of solitary researchers and rather frowned upon by universities, bestows upon it a new force and visibility, as well as an undeniable institutional presence. Indeed, which French university has not recently held a colloquium on "The Human and the Animal?"[9] Only fifteen years ago such a thing would have been unthinkable. In the early nineties, a general conference was held at the Sorbonne on animals in antiquity, but the topic was still relegated to the history of philosophy, in this case Greek.[10] While making an important contribution to the history of philosophy, this conference did not critique the status of the animal or philosophical discourse about the animal, which is an entirely different issue. A more telling example is the jury's topic selection for the 2012 statewide exam for students wanting to teach philosophy at the university level.[11] The fact that they chose the topic "the animal" is all the more remarkable in that the proposed topic, instead of reverting to the old phrase "man as rational animal," reflected a new phenomenological or etho-phenomenological understanding, asking exam candidates to expound on "the animal world." According to teachers of the last year of *lycée*, students are beginning to ask important questions about animals, and these questions are increasingly pointed, in part because the textbooks continue to treat "the animal" as a pseudo-Other deprived of everything with which man is endowed, so that each chapter begins by asserting with confidence that "man, unlike the animal" has a history, a language, consciousness, etc. This fact may have played into the choice of "the animal" as a topic for the state philosophy exam, but it seems clear that the intellectual climate has changed. Not that disciplinary practices and discourses have changed in and of themselves, but the question of the treatment of animals is increasingly being asked in the media; it sparks conversation, which in turn leads to personal insights, discovery of conviction, a sort of building up and fortifying of the issue. Common sense dares to rebel more convincingly against certain theories that seem too peremptory, too counterintuitive, too insufficiently proven, and that lack the ethical neutrality that characterizes philosophical exploration.

Animal philosophy, if this expression must be used, has, along with other disciplines, established itself at the cost of considerable effort. It has had to prove itself intellectually more than other fields have done; it has been forced to struggle to claim its respectability. Thinkers in this area have managed to elevate animals to the level of other major philosophical issues. Previously the source of much mockery, the animal question has become a major ontological, epistemological, moral, and political force; some even say that a profound anthropological shift is underway.

## QUESTIONING PATTERNS OF THOUGHT

### Continuist Perspective vs. Dualist Perspective: A Comparison

I would like to now advance a position that aims to free itself from the following two positions—radical human-animal dualism on the one hand, and human-animal continuism on the other—that have been wrongly presented, it would seem, as being mutually exclusive, with the latter supposedly prevailing over and defeating the former. Proponents of continuism like to say that "man is an animal," as if anyone still had serious doubts (we recognize that there may be exceptions), and they sometimes add, perhaps for the enjoyment of vexing those who uphold the idea of man's eminent metaphysical dignity, that he is an animal "like any other." Not only is this a shallow and hasty premise, but it also misses the point entirely; to strike a blow against the singularity of man, *le propre de l'homme*, it can find no better weapon than lowering humans to the same level as animals. This proposition is underpinned by the notion that "animality" is a sort of tabula rasa where everything is equalized; it denies the very idea of specificity. Man is not "an animal like any other" for the simple reason that the expression "animal like any other" has no meaning: *no animal is an animal like any other.* Each species has its own characteristics that form, piece by piece, a psycho-physiological whole. There are as many singularities as there are species. The almost exclusive interest paid to the human species should not include a reduction of all that is not human to a zero degree of

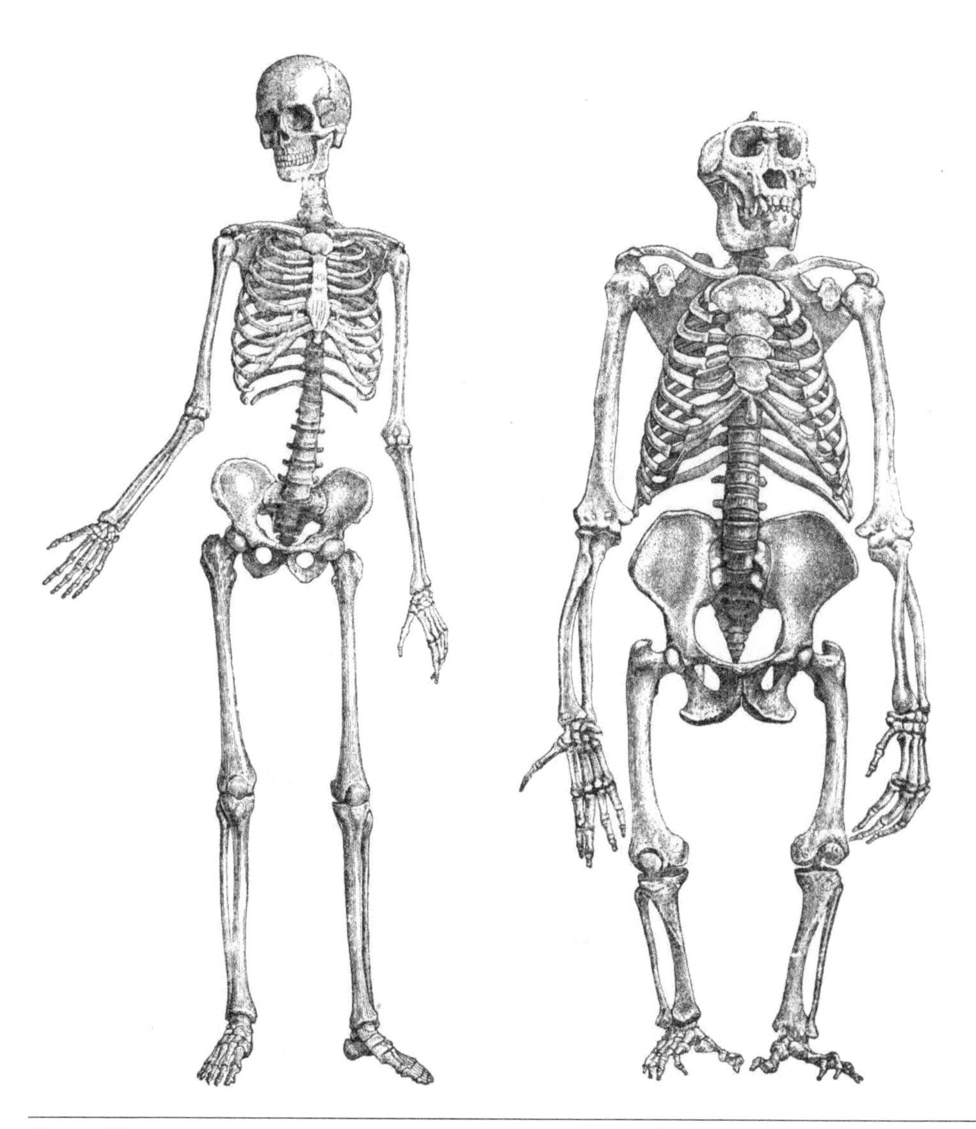

***Figure 1.*** "Skeleton of human and gorilla, unnaturally stretched." (Source: Brehms Tierleben, Small Edition 1927, Wikimedia Commons, http://en.m.wikipedia.org)

complexity. The fact that animals do not speak, at least in the sense that we understand language, is not a "lack" causing a loss of meaning, but a condition, a state of being, a *Stimmung* that indicates *another mode of existence.*

This all bears repeating: critiquing the continuist thesis does not mean removing man from his animal classification; nor does it mean proving man to be a metaphysical exception;[12] nor does it mean disembodying man, or casting him as some kind of dual being or "rational animal," that is, a body topped off with a mind or a consciousness. Rather, it means considering carefully and critically the *philosophical* beliefs about animals or animality to which continuism adheres. Thus it is not from the standpoint of evolutionary biology that the continuist position presents serious flaws, but as a *philosophical alternative* to dualism.

This continuism, such as it has been postulated by many philosophers (sometimes with the best intentions in the world), remains in effect anthropocentric; this is the reason for its philosophical flaws, not to mention its misguided nature. To continue thinking of animals while taking humans as the point of departure, without even questioning this assumption, leads inevitably to the following conclusions: that animals have a proto-culture, a proto-language, and something resembling a conscience, but are incomplete or undefined because their self-awareness is lacking, etc. In short, according to this perspective, the animal is always "lacking" something required to qualify as human. The "self-recognition mirror test" is considered to be decisive: whether or not an animal test subject wipes away a smear of lipstick applied to its forehead by a researcher is now supposed to determine the complicated issue of self-consciousness. It is unfortunate that psychologists and behavioral scientists who engage in this testing do not understand what phenomenologists call "consciousness of self" or "non-thetic consciousness of self," which is a consciousness that accompanies every experience—meaning that each experience of a behavioral subject is truly *personal*, that they are all first-person lived experiences. While it is certainly difficult to characterize *the* animal's consciousness, the key is to match consciousness with intentionality—in other words, to understand that animals participate in the constitution of a world, as evidenced by their consistently coherent and focused behavior. Animal consciousness, before we enter into other specific descriptions, could best be described as consciousness at the level of phenomena that is directed at objects that the consciousness in turns constitutes. And what is targeted by this consciousness cannot be conceived of without the "self" of the consciousness that is doing the targeting. Consciousness of self or non-thetic consciousness of self designates consciousness that carries the self of experience, but does not include the reflexivity that may eventually follow or accompany an experience, *re*present it or present it *once again* to the subject. If we wish to move towards a particularization of the consciousness of any animal, then we must turn to that animal's world, the specific world he has built for himself, something brought to light by Jakob von Uexküll—and what an epistemological revolution!

All things considered, animals emerge from continuism enhanced by a sort of pseudo-humanity or humanity in the making, but this making has no future. Animals would be replicas fashioned more or less after a human model that remains the culmination of all cognitive aptitudes, affective dispositions, and performances of any kind. And how could it be otherwise when animals are measured against humans?

This way of thinking constitutes a philosophical impasse and, as with radical dualism, we must break with its thesis. More specifically, we must break with the definition of a singular "animal," as before mentioned, as well as with a complicit approach that "measures" animals on the basis of human performance. Criticisms leveled at the Great Ape Project by some of the contributors of the present collection of articles pointed out that the granting of certain rights to great apes was based on their ability to demonstrate certain cognitive abilities. This is a step in the wrong direction. When one starts with the cognitive abilities of animals in order to conclude that they have attained (and indeed, can no longer "exceed," as opponents of animal rights do not hesitate to emphasize) the ability of a five-year-old human child, one essentially closes the door to all animals who do not reach this bar. This is the wrong approach for one basic reason: the moral criteria for basic fundamental rights is not the ability to count or to construct complex solutions, but an affective disposition towards oneself and the world, a phenomenological disposition, rooted in the connection between consciousness and the objects it forms, or what phenomenology calls experience.

Scientific research on the representational capacities of animals, whose academic rigor and seriousness we are not contesting, examines the behavior of animals in the laboratory through the use of laboratory material and equipment. Chimpanzees are placed in front of computers to determine their mental capacities, or they are subjected to surgical brain operations after which they are killed because they are no longer useful for testing.[13] This is *also* cognitive ethology. But this is not only what cognitive ethology is, and so we must immediately qualify this assertion by paying homage to the work of Donald Griffin on animal consciousness, as well as to ethology research done in the field by pioneers such as Jane Goodall, Mark Bekoff, or Frans de Waal, to mention only the best-known authors in France; their work does not fall within the scope of our critique. Based entirely on observations of the ways in which animals are and live their lives, these researchers seek to bring to light the life of the psyche. This life of the psyche should not be understood as a set of cognitive abilities distinct from behavior that remains solely physical; on the contrary, the life of the psyche is already part and parcel of such behavior. To quote a beautifully precise and succinct sentence from Husserl that should be committed to memory for the purpose of this thesis: "Behavior . . . appresents [presents in a direct-indirect manner] the psyche as its sign."[14]

## The Criteria of Lived Experience

To counter a way of thinking, we must modify its structure; it is not enough to check off the boxes (something that continuism, as we have defined it, does), because we must rethink the very nature of these boxes. Phenomenological approaches seem to offer the possibility of such a restructuring because of the primacy they give to lived experience (a tautological phrase in its own right, since all experience is by nature lived experience). The tools thus forged lay the groundwork for an ontology of animal life by showing that animals, far from being the simple creatures that philosophical tradition has confined them to being, live their lives in the first person. All of this amounts to a vast research program in which I have been involved for several years, and which I will sketch out briefly here.

Which living beings can be said to live their *own* lives, through a presence of self that is experience? Is it possible to determine the general tone of this experience? The philosophical task before us lies in working out two descriptions: one etho-phenomenological, which focuses on the content of the lived experience; the other more strictly philosophical, which focuses on the overall tone of this lived experience. By definition, lived experience has substance and depth, which is that of the life of the psyche, and not the supposed flatness of "regular life." In however inexplicit a lineage following Hegel, who attributes to animals "a sense of uncertainty, anxiety and unhappiness"[15] because of their deracination, Sigmund Freud, Kurt Goldstein, Gilbert Simondon, and before them Arthur Schopenhauer—each of course in very different ways—have all developed a theory of anxiety related to the condition of living that includes animal life and not just human life. It is clear that questioning the validity of the distinction between life and existence only makes sense if the definition of existence as determined by human standards is set aside in order to explore an understanding of existence as deeply rooted in vital structures. The question of animal existence can obviously never be investigated if existence is already unequivocally limited to humans. "By existence, we do not mean," writes Frederik Buytendijk, "the strict sense used by theories of being in current anthropology and ontology when referring to a mode of being reserved only for humans. For us, the term 'existence' designates a situation manifested by a behavior."[16] Concepts such as one's own body, flesh, organic subjectivity, or corporeal existence, which have been forged by phenomenological thinking, emphasize the dimension of life that is both embodied and transcendental. Behavior, as redefined by Maurice Merleau-Ponty in light of the concept of structure, forms the foundation of these concepts.[17]

These ideas, which I have developed extensively elsewhere, have led me to propose the following definition of animal existence. We speak of existence whenever a living being not only functions as a center from which to organize relations with and in an environment, but is also the subject of these experiences, or in other words lives them in the first person (the German term *Erlebnis* is clearer here because it refers to the lived aspect of an experience: "a lived experience"). Without referring to the notion of experience, existence loses its specificity and could be reduced to a third-person description of a course of action. To say that individual animals are subjects of *their own experiences* seems more appropriate than framing them as subjects of their lives, since the latter formulation might suggest that the subject is capable of surveying this life, of considering this life as a whole, while reviving the idea that such a subject should be able to project into the future, to make one or several "life plans."

We can see the idea of a biography beginning to make sense. By biography, we mean the tracing of a life as experienced by the subject himself: biography is lived experience perceived in the uniqueness of its trajectory; the character of being lived is what gives a life journey its sense of unity. As defined by Wilhelm Schapp, the concept of entanglement, which means the intertwining of things and beings to the point that it becomes impossible to separate them, gives the idea of biography profound depth. This entanglement includes temporal, spatial, physical, and cognitive horizons, but also the dream world (our days sometimes become entangled with our dreams): it is a supersaturated texture of threads that has no margins. To become entangled, a being does not need to recognize the concept of entanglement.

> A lion carries his history with him. He is old or young, domesticated or wild, healthy or sick, he also shows intents related to his imprisonment, as he sits behind iron bars . . . he appears as a story comprised of many intents, a biography, that is lost in the horizon. From this horizon fixed points arise. His age evokes his birth and, by the same token, his relationship with his parents. Certainly the umbilical cord that connected him to his mother is broken. But the connection itself remains, as the most intimate connection between two living beings . . . This connection extends backwards, forming a series that is impossible to take in at a single glance. Thus, an individual lion takes his place within the race of lions.[18]

***Figure 2.*** DESTA, LION FROM ABYSSINIE (ETHOPIA) IN CAPTIVITY IN THE JARDIN DES PLANTES, A GIFT FROM FRENCH PRESIDENT CARNOT. (SOURCE: HTTP://WWW.CPARAMA.COM)

Schapp adds that the place occupied by such a lion, just as the place occupied by lions as a species, is in no way comparable to a chance position in time and space: these singular lions appear in a certain way and in a certain order; they occupy one position and not some other. In this way, although all animals are excluded from formal history by virtue of being animals, their anonymous selves are nevertheless captured in a kind of history.

## FINAL REMARKS

How can this perspective that is open to the ideas of first-person experience, existence, and biography contribute to the foundations of an animal ethic that is the opposite of the ideology of "well-being" currently serving the interests of "animal production"? I am speaking here of the manner in which the food industry has appropriated the claim of "well-being" from animal-rights groups. While not far from a morality of compassion, utilitarian theories, and theories of animal rights, a phenomenology of first-person animal existence uses a different approach to give consistency and depth to what animals experience in life. More precisely, the development of the concept of animal existence leads to a critique of utilitarianism that merely condemns suffering without concern for the uniqueness of every life, of each life that was meant to be lived in and of itself, of a life experience that cannot be reduced to the absence of suffering; utilitarianism continues to view individuals as replaceable and interchangeable so long as the amount of happiness in the world remains the same. The "animal welfare" view, which consents to the continued exploitation of "animal resources" but instructs those who do so to cause no animal suffering—as if such a (false) hypothesis were even possible—is clearly unacceptable from a perspective that takes seriously the concept of animal existence, which holds to the principle that animals deserve to live their individual lives.

## NOTES

1. The expression *philosophie animale* is supported through its use in publishing: Vrin (a philosophical publisher if ever there was one) and Minuit (which publishes the review *Philosophie*) have adopted it. This immediately led to the use of another term, *éthique animale*, by the Presses Universitaires de France. In large part, the use of these expressions is due to two key authors: Jean-Baptiste Jeangène Vilmer and Hicham-Stéphane Afeissa. The influence of the English language is evident in the work of these philosophers, who present and analyze American moral philosophies on the status of animals. See for example Jean-Baptiste Jeangène Vilmer, *Ethique animale*, with preface by Peter Singer (Paris: PUF, 2008), and Jean-Baptiste Jeangène Vilmer and Hicham-Stéphane Afeissa, eds., *Philosophie animale: Différence, responsabilité, et communauté* (Paris: Vrin, 2010), with contributions from important Anglophone thinkers.
2. Jacques Derrida, *L'animal que donc je suis* (Paris: Galilée, 2006), 54; and *L'animal autobiographique* (Paris: Galilée, 1999), 282a.
3. Élisabeth de Fontenay, "La bête est sans raison," *Critique* 375–376 (1978): 707–729.
4. Élisabeth de Fontenay, *Le silence des bêtes: La philosophie à l'épreuve de l'animalité* (Paris: Fayard, 1997).
5. For some notable exceptions, see the work of sociologist Marion Vicart, *Des chiens auprès des hommes: Ou comment penser la présence des animaux en sciences sociales* (PhD diss., École des Hautes Études en sciences sociales, Paris, 2010), who has courageously opened a new chapter of study; and the work of historian Éric Baratay, who redefines history as a discipline in his book *Le point de vue animal: Une autre version de l'histoire* (Paris: Seuil, 2012).
6. *Animal Liberation* by Peter Singer was translated into French in 1993 by Louise Rousselle and David Olivier (Paris: Grasset) and has just been republished. *The Case for Animal Rights*, by Tom Regan, has just been translated into French by a moral philosophy scholar, Enrique Utria (Paris: Hermann, 2012).
7. Jean-Yves Goffi was the first French philosopher to present the ideas of Peter Singer, Tom Regan, and Joël Feinberg, with all the seriousness due these authors, in *Le philosophe et ses animaux* (Paris: Jacqueline Chambon, 1993) as well as in several articles. More recently, see Jean-Baptiste Jeangène Vilmer, *L'éthique animale*, and *Anthologie d'éthique animale: Apologies des bêtes* (Paris: PUF, 2011). These works deal with Anglo-American animal moral philosophy, or animal ethics. It is impossible to cite here all the works currently in the doctoral research stage that explore this moral philosophy.
8. This collective book was the subject of the debate between Paola Cavalieri and French philosophers that was published in the journal *Le Débat* 108, no. 1 (2001). The English-language book, edited by Paola Cavalieri and Peter Singer, was translated as *Le projet grands singes: L'égalité au-delà de l'humanité*, trans. Marc Rozenbaum (Nantes: OneVoice, 2003).
9. See Anne Simon's contribution to this volume for an enumeration of only some of these recent colloquia.
10. The proceedings of the conference were published under the direction of Barbara Cassin, Jean-Louis Labarièrre, and Gilbert Romeyer Dherbey as *L'animal dans l'antiquité* (Paris: Vrin, 1997).

11. Translator's note: *Agrégation* in the original French.
12. Philosopher Jean-Marie Schaeffer's critique of "the human exception" was not to everyone's liking (*La fin de l'exception humaine* [Paris: Gallimard, 2007]). See for example the response of philosopher and Jesuit Paul Valdier in his book *L'exception humaine* (Paris: Éditions du Cerf, 2011).
13. I feel compelled to mention an article that appeared in a collection on ethology and animal cognition; the article summarizes an experiment whose insignificant objectives and incredible cost astounded me. To study the "role of the anterior cingulate cortex in reward-based behavioral decisions among rhesus monkeys," Céline Amiez and Jean-Paul Joseph explain in a section entitled "Material and methods" that they placed the animal in "a chair for primates" (there is much one could say about the existence of such a chair) "facing a tactile slab." The animal had to perform certain "tasks." Some "surgical procedures had been performed," and the authors dwell at length on the substances injected to anesthetize the animal in order to affix to its head "using small stainless steel screws," a "steel bar," also stainless. An "acrylic assemblage" immobilized the head during this procedure. Then the authors list the painkillers that had been injected into the monkeys (two of them had been subjected to this procedure). In short, they treated the "material" in "ethical" fashion. This is what the authors wish to convey to us, as if the ethical question of animal testing could be reduced to the technical management of the material. Giving detailed descriptions of procedures is a recent phenomenon. Do they really believe this will answer a question that, in reality, deals with the very legitimacy of submitting animals to experimental testing, and not merely with the carrying out of these experiments? Instead of being reassuring, the ease with which the authors describe the "procedures" is extremely disquieting. This article, only one of many that could have been cited, perfectly illustrates that for these researchers who study cognition, animals are empty vessels (hence the apt title of the section "Material and methods") who can certainly feel pain, but who have no lived experience. See Céline Amiez and Jean-Paul Joseph, "Rôle du cortex cingulaire antérieur dans les choix comportementaux basés sur les récompenses," in *Autour de l'éthologie et de la cognition animale*, ed. Fabienne Delfour and Michel Jean Dubois (Paris: PUL, 2005), 35–47.
14. Edmund Husserl, *Méditations cartésiennes: Introduction à la phénoménologie*, trans. Gabrielle Peiffer and Emmanuel Levinas (Paris: Vrin, 1966), 184.
15. Hegel, *Encyclopédie des sciences philosophiques II: Philosophie de la nature* [1817, 1827, 1830], translated and edited by Bernard Bourgeois (Paris: Vrin, 2004), 324. For an overview of Hegel's philosophy of the animal as organism, see Florence Burgat, *Liberté et inquiétude de la vie animale* (Paris: Éditions Kimé, 2006), 187–205.
16. Frederik Buytendijk, *L'homme et l'animal: Essai de psychologie comparée* [1958], trans. Rémi Laureillard (Paris: Gallimard, 1965), 46.
17. Maurice Merleau-Ponty, *The Structure of Behavior*, trans. A. L. Fisher (Boston: Beacon Press, 1963). I have given a detailed account of Merleau-Ponty's philosophy of the animal question in Florence Burgat, *Une autre existence: La condition animale* (Paris: Albin Michel, 2012), 165–197.
18. Wilhelm Schapp, *Empêtrés dans des histoires: L'être de l'homme et de la chose* [1983], trans. Jean Greisch (Paris: Éditions du Cerf, 1992), 81–82.

# Like the Fingers of the Hand

## *Thinking the Human in the Texture of Animality*

DOMINIQUE LESTEL

*Translated by* MATTHEW CHRULEW AND JEFFREY BUSSOLINI

European thought has traditionally addressed the question of animality in terms of a hygienic border, the problem being how best to characterize what distinguishes humans from animals—that is to say the *propre de l'homme* or that which is "proper to the human"—namely, a characteristic that humans alone possess and that so differentiates them from other animals that it pushes them beyond animality. Such a notion is highly problematic. Searching for competencies that one would find only in *Homo sapiens* is a more reasonable project, on the condition, however, of being sensitive to the pitfalls of the concepts mobilized and to the vicissitudes of reasoning that one would be tempted to use to get there. To think animality today, we cannot do without a ferocious deconstruction of contemporary comparative psychology and ethology. Since the end of the second half of the twentieth century, a growing number of theorists and activists of the animal cause have sought to give to humans and animals a similar legal status. I myself defend an even more radical posture, the premises of which can be found in Paul Shepard's work: humans constitute themselves as human in the very *texture*[1] of animality, and human/animal interpenetrations do not belong to a more or less distant past but are still always present, even if human/machine connections have now entered into competition with them.

### THE MAJOR POSTURES OF THE PHILOSOPHY OF THE ANIMAL

In Western thought,[2] the philosophy of the animal is expressed broadly through six major postures from which all the others follow. (1) *The posture of the animal machine* is the most widespread in the animal sciences. It holds that the animal is a transparent causal machine for which it is possible to provide a comprehensive instruction manual.[3] (2) *The posture of the animal victim* instead emphasizes animal rights and engages in various practices of social protest. Peter Singer is representative of it. This very Western-centric vision of the "sad beast" and the "animal victim" is that of an innocent victim who must be protected by humans.[4] (3) *The posture of the inferior animal* is that of the "poor animal," which can be exemplified by Martin Heidegger, but its premises date back to Ancient Greece. The animal is characterized by its supposed "lacks" when compared to humans. (4) *The posture of the mysterious animal* is that found in Adolf Portmann, Hans Jonas, and Jakob von Uexküll. For the latter, for example, the animal is "mysterious" because she interprets the world around her through physiological senses she does not share with us. (5) *The onto-evolutionary posture* is discussed by the American Paul Shepard, who is its most interesting adherent, but can also be found to some extent in the very particular "monism" of the French philosopher Raymond Ruyer, for whom seeking the difference between human and animal is simply meaningless, humans being predominantly composed of bacteria. In this posture, the relationship between humans and other animals is constitutive of human identity itself: humans have become human through their proximities with animality and not in distancing themselves from it. (6) *The hermeneutico-interactionist posture*, finally, for which the human/animal relationship determines what each of them is, and in which one can place Jacques Derrida and Vinciane Despret, even if each occupies this position in a very different manner—Derrida from a praxiology of everyday life,[5] Despret from that which she calls a bit curiously an "ethology of ethologists."

## REESTABLISHING A CONCRETE AND CONSTRUCTIVIST PHENOMENOLOGY OF THE ANIMAL

At the interface of the onto-evolutionary posture and the hermeneutico-interactionist posture, my approach to animality falls within the framework of a concrete and constructivist phenomenology that cannot be reduced to the legitimacy of animal rights or to an ethics of human/animal relations. Four major elements characterize this phenomenology: (1) It is based first of all on a critical epistemology of ethology and comparative psychology. It also takes the form of an "epistemology of rupture" and is therefore opposed to the "epistemology of complicity" that is usually adopted by the philosophy of science. (2) It adopts an ultra-continuist position that considers that all human/animal differences are secondary compared to their mutual convergences and to what I designate by the term *inter-texturity* [*inter-texturité*] with each other that results from them. The anatomical metaphor is illuminating here: to seek the specificity of humans in relation to other animals makes no more sense than seeking that of the heart in relation to other organs. We find in both cases a difference that does not divide, but rather allows a life in common that far exceeds mere cohabitation. (3) It reckons that the question of animality, in the twenty-first century, is no longer only that of "natural" animality, but also that of "artificial" animality (hybrid creatures, genetically modified organisms, animalized artifacts) from which it has become inseparable. (4) It notes, finally, that the Western intellectual tradition has significantly damaged its intellectual resources for thinking animality, and that philosophy must be open to other forms of thought to reestablish this theme of animality in a fruitful way. Obsessing about human/animal difference has profoundly handicapped the Westerner for many centuries and has limited the capacity for thinking about human/animal overflows [*débordements*]. In the pages that follow, I begin by showing why it is essential to refer to the animal sciences to construct a philosophy of animality, and why these animal sciences need to be deconstructed (and reconstructed) by the philosopher—and not used as such. I mention in particular the need to elaborate an ontology of position, devoid of all substantialist ontology.[6] I then specify my hyper-continuist position—without, however, denying the unique place that the human has *culturally* acquired in the sphere of the living. Inversely, if the human is not only animal, other animals are no less so! Hence the need to rethink human obligations vis-à-vis other animals and my proposal to do so in terms of ecological obligation rather than in terms of moral obligation: a notion of ecological obligation the meaning of which I will explain using the concepts of infinite debt and the ethics of reciprocity. I conclude with the need to open Western philosophy up to novel forms of thought to better understand the animal, and with the need to think about the openings to which contemporary artificial animality gives rise.

## THE PROBLEM OF THE ANIMAL SCIENCES

Philosophical thought about the animal too often neglects the question of our scholarly knowledge of the animal, except to lay claim to it in an imprudent and rather rushed manner when these results go in the direction desired by the one who mobilizes them. However, these sciences have always played an important role in anti-animal dispositives developed in the West, as though animal sciences must necessarily take the role of witness for the prosecution. Ethology and comparative psychology indeed follow a largely unspoken but terribly restrictive agenda that consists of wanting to establish scientifically that which is "proper to the human." Deconstructing these sciences is therefore necessary. Such a task is, however, far from simple because the difficulties are at once conceptual, theoretical, and empirical. Ethologists and psychologists have a naive relationship to the concepts used and arguments mobilized. This is moreover equally true of a somewhat marginalized innovator such as Jane Goodall as it is of a more traditional university mandarin such as her mentor Robert Hinde. The ethologist is all too often in a situation similar to that in which one of the fingers of a hand would try to give a representation of itself that would be independent of not only the other fingers but also the hand, while claiming that it is *obviously* part of the hand. One does not observe rationality, for example, as one observes predatory behavior.[7] Ethologists and psychologists seldom realize that the concepts mobilized in ethology have no natural necessity but have acquired their meaning in the context of that which is proper to the human, and in a long and tumultuous cultural history. These researchers operate in a realist-Cartesian paradigm that prevents them from asking in an effective manner some of the most interesting questions in the field, and other paradigms are undoubtedly more fruitful.[8] For the bi-constructivist paradigm, for example, the ethologist must invent ways to account for the ways in which the animal invents its relations with the world. Ethology and comparative psychology are constructed in such a way that it is simply not possible to give a positive response to some of the most unsettling questions one might ask about animal intelligence—for example, about artistic practices or the relation to death and corpses. The question of animal art cannot, for instance, be seriously discussed by researchers who refuse to give the slightest relevance to singular phenomena, who immediately disqualify any reference to a non-behaviorist concept of signification,[9] and who obsessively and pathologically reject anything that resembles anthropomorphism.[10]

## CONSIDERING THE PERSON FROM THE PERSPECTIVE OF A RELATIONAL ONTOLOGY

Nothing permits excluding the animal from the space of the subject that the Westerner has claimed in a privileged or even exclusive fashion. Nor does anything prevent thinking that the question is ultimately more complex than having to decide in a dichotomizing way if the animal is a subject or not. To consider the animal as a machine is in any case pure nonsense that is incompatible with the available empirical data and is often based on dubious reasoning. Every animal is on the contrary a *subject* (in the sense that each animal is an interpreter of meaning); some are *individuals*, and certain animals become *persons* through interaction with humans.[11] An agent is not a subject because he *possesses* the required competencies. Something like a "person" does not exist objectively. Every person is a character who plays a role in a relational narrative process.[12] The relevant question is not therefore whether this or that agent is truly a person or not, but to what extent some of us can comport with her as if she is, which is a very different question. In other words, our relations to an animal are never purely causal: they are still largely semiotic and narrative and have no meaning except in the stories through which we create our identities and those of the agents in question. Such an approach begins with relationships rather than with an essence of the actors; one searches in vain for universal metanarratives that would validate or refute the multiplicity of small stories through which we make sense of others—whether human, animal, artificial, etc. There is no ontology that is not already relational and contextual. The right question is not whether animals are people, but whether we are prepared to live in a society in which animals could *legitimately* be considered as persons. In this constructivist perspective, literature and art clearly have a major role to play in a dynamic in which we produce the real more than we generally notice. A thinker such as Paul Shepard goes even further in considering that a human cannot be a person unless at least some animals are.[13] We could consider the opposable thumb as that which established human difference. But if that famous thumb has no other fingers to which it can be opposed, humans can only suck it like a baby, whining and cursing their fate.

## THE MESHES OF EXISTENCE

The foundations of a thought devoid of all substantialist ontology are expressed in an elegant way with the metaphor of the meshwork [*maillage*] through which each agent (human or other-than-human) is knitted. Tim Ingold rightly distinguishes the *network* from the *meshwork*, and assigns a greater relevance to the second than to the first.[14] The network is abstract, the meshwork concrete. The meshwork acquires its consistency through the practice of the heterogeneous assemblages from which it is constituted, and not on the abstract logic that constitutes networks. For Ingold, the organism is not a closed entity surrounded by an environment, but a point of convergence of open lines that interweave into each other in a fluid space, which constitutes at the same time their own environment.[15] An agent is given form as a weft of the meshes within a hybrid community,[16] and this ontological ball of string is determined by the assemblages that it builds with other agents, be they human or other-than-human.[17]

## AN ULTRA-CONTINUIST POSITION

The absence of substantialist ontology makes it difficult to distinguish sharply between human organisms and other-than-human organisms. The discontinuist posture that relies on the concept of the "proper to the human" concerns itself not with what distinguishes humans from other animals, but rather with what distinguishes humans so much that it pushes them beyond animality. Numerous animal species, however, have characteristics that are not found in any other species, without our evoking about them any kind of "proper" status that would place them apart. The comparison between human and animal is too often undertaken between problematic data about the animal and platitudes about the human. Certainly, nothing serious can result from this. Philosophers who are interested in these questions must not only educate themselves in ethology; they must themselves practice this science in a critical way. They must slip into the skin of an especially critical ethologist, who would, as a consequence, no longer exactly be an ethologist.[18] They must also have recourse to appropriate arguments. Most of the arguments that are called upon to consider a topic such as that which is proper to the human are simply inadmissible. What is at stake here are in fact *differences of differences* and not *differences of competencies* (themselves very problematic in the sense that ethologists blithely confuse competency and performance). The thumb is very

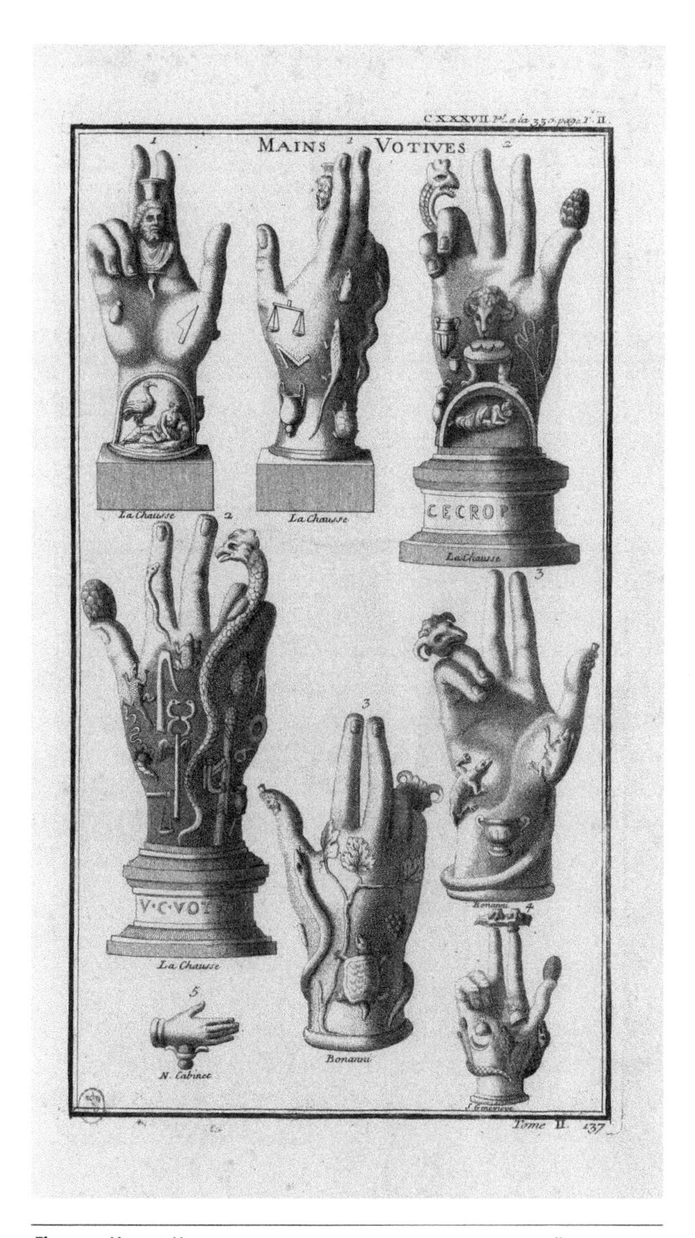


***Figure 1:*** VOTIVE HANDS, ILLUSTRATION OF CHAPTER ON HANDS "HONORED AS DIVINITIES" IN ANTIQUITY, IN *L'ANTIQUITÉ EXPLIQUÉE ET REPRÉSENTÉE EN FIGURES*, BY BERNARD DE MONTFAUCON, 1719, VOL. 2, FIG. 137 FACING P. 330. (SOURCE: BIBLIOTHÈQUE NATIONALE DE FRANCE, USED WITH PERMISSION)

different from the little finger, but to make it an entity that is ontologically distinct from the hand would be a mistake. As part of the same hand, and despite their real differences, thumb and little finger belong to the same category—even if such a realization hurts the vanity of the thumb. Many animals are different from other animals; is this difference similar to or different from the difference of humans from other animals?

## A POSITION OF CULTURAL DOMINANCE

The human can thus be a very different animal from other animals without this difference, however, nullifying its intrinsic animality. The weakness of the discontinuists who always put forward differences is that they are unable to explain rigorously why *a* difference makes *the* difference, and starting from when such a rupture occurs. To object that the destiny of humanity is still very different from any other species, and that humans have a greater influence on the world than any other species, is both false and true. False, because we must acknowledge that species such as bacteria have already had a much greater influence than humans on land ecosystems, or that the first fish to leave water (the *Tiktaalik rosae*), from which derive all terrestrial vertebrates, still has had an influence that the human does not equal or even approximate! Humans have today acquired a privileged position in the world, and it would be ridiculous to deny it. However, this preeminence has been acquired as a result of great cultural feats that do not justify assigning a particular ontological status to the human species, but that lead only to recognizing an exceptional capacity to culturally transform the environment to their advantage. The discontinuist position tries to protect the human, and it is important to understand this profound fear. But it is instead by acknowledging the complexity of the animal that we can substantially increase the density of the human.

## ANIMALITY OF THE HUMAN AND ANIMALITY OF THE ANIMAL

To think the human *as* animal is already out of reach for the majority of humanist thinkers, but it is necessary to be bolder still. The difficulty of thinking the human *in* the animal is correlative to the difficulty of thinking the animal as animal in the Western intellectual tradition—to do so otherwise than as that against which the human is defined, and to not only do so from a strict biological point of view. It is in this regard disturbing that one still evokes the "animality of the human," as if this expression were obvious, and that one never speaks of the "animality of the animal," as if such an omission did not pose a problem. I retort that of course the human is certainly an animal, and that *it* is nothing but that. But why would the animal be *merely* animal? An animal is always more than an animal. There is always a *surplus* or *excess* to animality in the very fact of being animal since every particular animal has a singular way of being—no animal having, for example, exactly the same biographical trajectory as another. Far from breaking from this fundamental rule of animality, humans simply follow it in their own cultural way. Adopting the opposite of the traditional position by believing that the animal is a human like everyone else (or almost) prolongs de facto a tradition that refuses to truly ask what an animal is rather than what said tradition is opposed to. The thumb is actually very different from the other fingers, but each finger is different from the others, and these differences are best expressed in the organization of the hand and in their belonging to the same biological texture. The fingers are never as similar to each other as when they have been cut and arranged in a line one beside the others. Such an exhibition would obviously have nothing to do with a theory of the hand; it would fall instead under a psychopathology in need of emergency treatment.

## MORAL DUTY VS. ECOLOGICAL OBLIGATION

If the human is an animal like any other (that is to say, as different from other animals as other animals can themselves be from other animals), his duties vis-à-vis other species are not like those of other animals, at least in the sense that animal-rights advocates usually express. In other words, the human has no more moral obligation vis-à-vis other animals than other animals themselves have vis-à-vis the biosphere. It is not a matter of believing that humans can do what they want, but of acknowledging that the constraints they impose on themselves vis-à-vis other animals do not fall within morality. The obligation of humans vis-à-vis other animal beings is rather an *ecological* obligation that refers to the need to maintain an ecosystem *viable for humans* (whatever this concept would mean otherwise) and to the symbolic obligations that help determine their identity; it is thus not a moral obligation that requires respecting prohibitions based on a priori representations of Good and Evil. The ecology that interests me here is, however, less that of ecologists (based on trophic chains, subsistence niches, and reproductive strategies) than an *ecology of identity* such as we find discussed by Paul Shepard. Such a conception of ecology in the first (or second or third) person does not necessarily have the mystical connotations that might be too easily attributed to it; rather it refers to the idea that to be a human is to take the form of an existence that cannot be such except by feeding on the existences of other living beings and in allowing other beings to feed on it. Such an ecology of identity can be comprehended more precisely through concepts yet to be developed—such as that of infinite debt (that humans incur vis-à-vis animality) or the obligations of reciprocity that jointly maintain humans and animals, but also plants, mushrooms, noncorporeal intelligences,[19] etc.[20]

## THE INFINITE DEBT

Philosophical thinking about the concept of debt has been singularly renewed in contemporary French thought,[21] but it has never been applied to the animal. This human debt towards the animal is infinite because it is immersed in the immemorial human past, because it extends to the furthest human future, and because it has a depth that is impossible to reach. This infinite debt is from species (*Homo sapiens*) to species (all the others), from cultures to ecosystems, but also from me to those animals who dwell in my living space as well as in my ontological sphere. It is because I became human over hundreds of thousands of years through my assemblages with the animal that I have an *infinite* and *personal* debt towards the animals who share my life. This debt is therefore a species debt that I inherit personally. It constitutes a major existential and ontological force, because it is largely through this debt that I constitute myself as human. It also represents an important spiritual and existential vulnerability: I am only human to the extent that I am the host of the animality that dwells within me. The debt that the human has contracted with the animal reaches an unattainable depth because it is metaphysical and constitutive of the human itself. This debt is infinite, because the human can never repay it satisfactorily. There is no point at which someone (whoever they may be) can claim that the debt is repaid, because a human being is inherently in debt vis-à-vis other living beings, particularly animals. Humans must therefore organize their lives so as to continually repay a debt that is impossible to repay by its very constitution. Such a situation is unbearable for the Westerner who always judges dependence negatively but who must recognize that he is nothing without this animality, which he nonetheless judges to be so inferior to him. This debt does not, however, relate to the economy except in a superficial way. It should rather be conceived as that which a child can have vis-à-vis the efforts and sacrifices that her parents made for her, as she becomes a full-fledged being (which does not mean she became what they wanted her to become). The human was born by chance, incidentally fulfilling a *promise* (not *the* promise) that animality carried in it. Before the disappearance of the dinosaurs, it was not *Homo sapiens* who had the highest probability of emerging. A species of superintelligent bird was more likely. By removing all animality, a transhumanism like that of Nick Bostrom or Ray Kurzweil finds a way, not to repay the debt, which is impossible by its very nature, but to disqualify and cancel it. This techno-utopian project retrospectively illuminates the hidden humanist project that it completes: to rid humanity of his animality, that is to say, to rid humanity of itself. Such an objective illuminates a major constant of humanism in European intellectual and moral history—its hatred vis-à-vis the animal and animality that leads to factory farms, the systematic destruction of ecosystems, and animal experimentation as it is practiced today.

## AN ETHIC OF RECIPROCITY

The notion of reciprocity is as neglected as that of infinite debt in debates on shared life between humans and animals. The discussions about animal experimentation remain very unsatisfactory because they oppose two irreducible positions that are rooted in nonnegotiable moral imperatives. Those in favor emphasize the need to save human lives while those against object to the suffering of the animals involved. An ethic of reciprocity considers, on the contrary, that *everything* is negotiable on the condition of being willing to provide the means. The fundamental principle of a relational ethic of reciprocity is not that one should never take, but that one can only take in accordance with what one gives. This results in a bioethics that is as much *prescriptive* (about what I must give considering what I want to take) as *proscriptive* (which forbids me from doing that which does not enter into a relationship of fair exchange). More than a set of morals, it is an ecology of reciprocity and an ecosystem of interests to guide my actions vis-à-vis other living beings. This means we must be prepared to recognize that the human does not have absolute priority over animals, and that conversely animal suffering may sometimes be entirely acceptable. Such an approach may undoubtedly offend some, but it is not new. Among the Confucians, for example, reciprocity is not modeled on a simple relationship of give and take, but on the interconnection and interdependence of those involved in the negotiation.[22] In other words, the relationship of reciprocity commits the researcher to the animal *with* whom she experiments, meaning she must invent forms of life in common in hybrid communities. This notion of reciprocity is far from simple; for example, how could the human be at the same time judge of and party to the relationship with animals on whom she experiments? The objection is valid, but it is not decisive. The researcher, first of all, can be sensitive to the ethical dimension of their work. After all, most parents are not abusive in the education of their young children, even if the children take part only superficially in the choices made about them. Secondly, laboratory animals may benefit from other allies—for example, activist groups for the protection of animals—that can influence the way in which research is conducted.

An anthropologist of fingers who specialized in the management of fingers in different cultures[23] would no doubt be very sensitive to the subtle interaction of fingers with each other in certain complex social situations, such as those in which two lovers hold hands. She would describe in detail the major role of thumbs in this loving exchange and might analyze this in terms of the primary finger that catalyzes the reciprocal caressing of the other fingers, but that does so without trying to take the upper hand.

## OPENING PHILOSOPHY UP TO UNEXPECTED WAYS OF THINKING

The Western intellectual tradition is profoundly zoophobic, but few are those who wonder if it is still possible to think animality in such an unfavorable cultural context.[24] Finding a way out of an intellectual space that is so negatively overdetermined in relation to animality has, however, become imperative. The difficulties encountered create the opportunity for transforming the philosophical framework classically used, opening it up to non-Western intellectual traditions and even to practices of thought that differ substantially from the usual canons of writing and philosophy. The academic division of knowledge subtly neutralizes all truly different ways of thinking about the animal and animality. Some theorists of animal liberation take the counterpoint to the usual Western postures by positing that an animal is a human like any other. Such an approach, however, is as unsatisfactory as that which would advocate general debauchery as a response to Puritanism. It is more fruitful to invent original postures that escape the great dichotomies that continue to structure European thought. To believe that we could *rearrange* Western thought that is rooted in the human-animal opposition in such a way as to create a space for the animal is an illusory hope. Philosophy has largely abraded its potential in gradually imposing conceptual constraints that block it considerably, or even sterilize it.[25] Today, we can consider the question of the animal as testing philosophy in a general way: how far is it willing to go to truly think animality? How far is it prepared to decolonize itself—that is to say, to abandon the dualisms inherited from Greek thought?

A number of options have yet to be seriously explored. The relational animist approaches to ontology described by Nurit Bird-David,[26] Eduardo Viveiros de Castro, and Graham Harvey[27] invite Western thinkers to truly open up to perspectives that are still very foreign to them. Animist systems offer valuable resources for opening philosophy up to conceptual adventures with which it is unfamiliar. These are obviously not models to follow blindly, but rather relevant examples for testing more concretely Western philosophy's capacity to open up. Animist beliefs and institutions can help philosophy to think animality differently: not by *explaining* these systems (which would prolong the colonialist illusion under which too many European thinkers still fall), but by allowing itself to be *intoxicated*[28] by the animist approach. Such a gamble is obviously not without risk, but it is illusory to believe that anyone could today think about animality without taking any risks—rather, it is because it provides a real challenge for thinking in general that animality is so interesting today. From this perspective, the desire to reclaim such an intellectual tradition is a real challenge that it is exciting to want to meet. But it is precisely because animism is irreducible to most of contemporary philosophical thought, and therefore so shocking to it, that it is so interesting today. What is at stake is not a return to a more or less imaginary animism, but rather a reinvention of the meaning that animism might have in a hyper-technological age like today, and a restoration of the relevance of its practices. It may in the end be the only chance we still have to save animality—and us with it.

## ARTIFICIAL ANIMALITY AND POST-ANIMALITY

The position I defend attempts, moreover, to think the major changes in animality that we face today. To my knowledge, no one has considered that the very phenomenon of animality might undergo a change in history.[29] Animality is always thought of as an invariant essence,[30] from the Paleolithic to the twenty-first century, even if some animals may change, for example through domestication. We have in any case great difficulty in thinking together about our phylogenetic and cultural histories, and we believe perhaps that such complication is not necessary. A static vision of animality is, however, largely problematic. The spectacular progress of NBTIC technologies (Nano, Bio, Information and Cognition Technologies) constitutes a turning point in this regard. Hybridizations and genetic or metabolic manipulations already have a potential that most of us are hard put to imagine. In 2011, for example, the UK Academy of Medical Sciences proposed to ban three types of scientific practice: grafting neurons onto monkeys so as to give them cognitive skills that should remain proper to humans such as language; manipulating germ cells in such a way that human/ape hybrids are viable; and giving animals a face that resembles that of a human.[31] Legislators rarely try to ban what is only fiction. But post-animality is already very present today under quite varied aspects that exceed biotechnologies. The spectacular development of animalized artifacts (such as Tamagotchi or Sony's AIBO) opens up very new spaces (both positive and negative) that one should think about seriously, and "talking" animals (primates like Kanzi or parrots like Alex) engage us in very unsettling and unprecedented spaces,[32] in particular because the point of view of the animal suddenly appears to *us* as very real. One could give many more examples. Humans today are at a turning point in their existence; whereas they have always lived with two hands consisting of five fingers (at least for the vast majority of human history), they realize all of a sudden that they have access to a multiplicity of hands that themselves have a multiplicity of fingers. The pretention of the thumb to have a special status, already ridiculous, becomes [even] more comical. In this context, the common expression "to twiddle one's thumbs" has probably never been more relevant.

Humans have never been able to contain animality, but we subject ourselves to it in ways that transform over time. Today we undoubtedly live in a situation as revolutionary as that which made us pass from the Paleolithic to the Neolithic era, when we began to domesticate animals and cultivate plants. Just as we have forgotten how animals contribute to making us human, and we fail to understand how our humanity is constituted with them, we have the greatest difficulty in understanding exactly how the new animality substantially reshapes what it means to be human.

The question of the *intermediate* level of our becoming-human has taken on an unprecedented relevance of a singular timeliness. At the "macro" level, we have been very attentive to the influence of culture on becoming-human—this is the problem at the heart of classical philosophical anthropology. At the "micro" level, contemporary biology has made spectacular advances in which are involved, though in a somewhat hasty way, many specialists in the neurosciences. The intermediate level between the biological and the cultural, that of "life in common" with other living beings, remains on the contrary still singularly neglected despite its importance. The preference of a growing number of people for animalized artifacts compared to natural animals serves to remind us that being human remains a very unstable position, and we have not yet reckoned the most dramatic consequences of our suspicious attraction to these strange, partially autonomous machines.[33]

## CONCLUSION

In the twenty-first century, the very theological question of clean borders between human and animal has become secondary; that of convergences between them, and of the lives in common[34] that result, has instead taken on a burning timeliness. Restricting such proximities to a morality, an ethic, or a right of the animal does not allow us to take stock of what is at stake. Such a reduction neutralizes further the challenge posed by the animal to the contemporary West. The human is nothing without the animal, and a major question of contemporary philosophy is to try to understand what this intuition truly means—and to do so before the last animal has definitively disappeared. In this text, I have defended an extreme position, namely, that the human is constituted not against the animal, nor even with it, but in it, at the interface of a hermeneutico-interactionist approach and an onto-evolutionist approach. This approach to animality is part of a concrete phenomenology that is at once evolutionist, hermeneutic, interactionist, and animist—and that fully assumes its speculative dimension. In an era that favors conceptual austerity, critical prudence, and positivist realism, such a posture will undoubtedly be seen as a form of provocation, but the joyful imagination must always remain a major resource for the philosopher. I understand that certain thinkers want to go on a dry diet, but we have no reason to make analytical anorexia (or conceptual obesity) into major virtues. Being human has always meant living with other-than-human agents: animals, plants, fungi, noncorporeal intelligences, and now machines equipped with some initiative and an undeniable capacity to resist what we want to do with them.

## NOTES

I would like to thank Louisa Mackenzie, Stephanie Posthumus, Hollis Taylor, and Jeffrey Bussolini for their constructive remarks on previous versions of this article.

1. The reference to the notion of *texture* signifies that the human/animal relation is not only social but also cognitive and above all metabolic.
2. It is a bit premature to speak of "Western culture" in a general manner, as if such an entity really existed. As an antidote, we can read Tim Ingold, *Companion Encyclopedia of Anthropology*

(London: Routledge, 2002), xiii. It remains nonetheless true that each culture has characteristics that distinguish it from others.

3. Richard Dawkins argues that every animal is a megamachine composed of a multiplicity of micromachines, in *The Selfish Gene* (Oxford: Oxford University Press, 1976), 2.
4. Translators' note: It bears noting that here, in addition to the French *animal*, Lestel uses the term *bête*, which means both "animal" and "stupid," thus highlighting an important dimension of anthropocentrism deeply sedimented in language. Jacques Derrida and others have considered this multivalence of *bête* at length. It should also be noted that the word *bête* can be used as a term of endearment or familiarity; for example, a farmer might say, when bringing in the cows at night, "il faut rentrer les bêtes."
5. One of the bizarre aspects of Derrida's position is that his interest in the animal never led him to be seriously interested in ethology or in comparative psychology.
6. A substantialist ontology supposes that there are beings in themselves, independent of the relations that they can undertake with other beings, and that these beings exist independently of the effects that these relations can have.
7. See Dominique Lestel, "What Does It Mean to Observe Rationality?" in *Rational Animals, Irrational Humans*, ed. Shigeru Watanabe et al. (Tokyo: Keio University Press, 2009), 44–66.
8. See Dominique Lestel, "What Capabilities for the Animal?" *Biosemiotics* 4 (2011): 83–102.
9. For a recent article representative of this situation, see A. Lerch, P. Roy, and F. Pachet, "Closed-loop Bird-computer Interactions: A New Method to Study the Role of Bird Calls," *Animal Cognition* 14 (2011): 203–211. The article "shows" that "this experiment is the first illustration of how closed-loop bird-computer interaction can be used productively to study social relationships."
10. Anthropomorphism is seen in ethology as a major methodological error when it should instead be seen as a moral taboo that reinforces the human/animal boundary. Frans de Waal showed precisely and rigorously how refusing every form of anthropomorphism is as problematic as accepting them all, in *The Ape and the Sushi Master* (New York: Basic Books, 2002). I discuss this question in detail in "Non-Human Artistic Practices: A Challenge to the Social Sciences of the Future," *Social Sciences Information* 50 (2011): 3–4, 505–512, and in "Could Beethoven Have Been a Bird and Could Picasso Have Been a Fish? Philosophical Problems of an Ethology of Art," in *Logic and Sensibility*, ed. Shigaru Watanabe (Tokyo: Keio University Press, 2012), 171–181.
11. See Dominique Lestel, *L'animal singulier* (Paris: Seuil, 2004).
12. See Dominique Lestel, "Portrait de la personne comme personnage," in *Personne/Personnage*, ed. Thierry Lenain and Aline Wiame (Paris: Vrin, 2011), 123–137.
13. See for example Paul Shepard, *Thinking Animals* (Athens: University of Georgia Press, 1978).
14. Tim Ingold, *Lines: A Brief History* (London: Routledge, 2007). The meshwork/network opposition is a major point of Ingold's critique of Bruno Latour.
15. Tim Ingold, "Point, Line and Counterpoint: From Environment to Fluid Space," in *Neurobiology of 'Umwelt': How Living Beings Perceive the World*, ed. Alain Berthoz and Yves Christen (Berlin: Springer-Verlag, 2009), 141–155.
16. On the notion of hybrid community, see Dominique Lestel, *L'animalité* (Paris: Hatier, 1996).
17. The term "other-than-human" is from Alfred Irving Hallowell, "Ojibwa Ontology, Behavior, and Worldview," in *Culture in History: Essay in Honor of Paul Radin*, ed. Stanley Diamond (New York: Octagon Books, 1960), 17–49. This expression seems more accurate to me than that of the "non-human" because it avoids including the nonpersonalized objects that are increasingly included among the "non-human," thus neutralizing any subversive charge.
18. Ethology rests on fundamental postulates that are all the more difficult to challenge because

the majority of ethologists *really believe them to be*. Within this paradigm, all behavior must be explained in terms of the interest that it presents from an evolutionary point of view, and concepts such as "comfort" or "signification" have only an impoverished meaning. The situation has deteriorated over the last few decades. From the fifties to the seventies, well-known and respected researchers like Frederik Buytendijk or Georges Thinès could still use a phenomenological point of view to study the animal, which would be impossible today.

19. The term is a little strange, but the more common term "spirit" is less adequate. Contemporary Western intellectual culture holds that such entities do not exist. *All* other cultures think the opposite. Deciding prematurely and too abruptly results from colonial ethnocentrism (I know because I have the science [and the canons]). On the empirical level, a number of people, even in the West, modify their behavior, beliefs, etc., as a function of such entities. From the point of view of a radical empiricism such as that of William James, one must grant such experiences a certain relevance in furthering the understanding of human behavior. In keeping with James, one could even say that in certain situations, noncorporeal intelligences and humans are "co-implicated." See lesson 2 of *A Pluralistic Universe* (London: Longmans, Green and Co., 1909).
20. For a more in-depth discussion of this infinite debt and ethics of reciprocity, see Dominique Lestel, *L'animal est l'avenir de l'homme* (Paris: Fayard, 2010), chaps. 3 and 4.
21. In particular see Marcel Mauss, "Essai sur le don: Forme et raison de l'échange dans les sociétés archaïques" (originally published in *Année Sociologique*, 1925), and Jacques Derrida's reading of Mauss in *Donner le temps* (Paris: Galilée, 1991) and *Donner la mort* (Paris: Galilée, 1992).
22. The work of Julia Tao at the University of Hong Kong and Roger Ames at the University of Hawai'i is important in this respect.
23. The Japanese novelist Yoko Ogawa's strange 1991 novel *Yoaku no ai* (Tokyo: Fukutake Shoten) presents a character who could be such an anthropologist.
24. See, however, Val Plumwood, "Nature in the Active Voice," *Australian Humanities Review* 46 (2009): 113–129.
25. On this point, see François Jullien, *Un sage est sans idée* (Paris: Seuil, 1998), especially chap. 7.
26. Nurit Bird-David, "Animism Revisited," *Current Anthropology* 40 (1999): 67–91.
27. Graham Harvey, "Guesthood as Ethical Decolonising Research Method," *Numen* 50 (2003): 125–146.
28. In philosophy, the concept of intoxication is mobilized first and foremost by Avital Ronell, *American philo: Entretiens avec Anne Dufourmantelle* (Paris: Stock, 2006), 23–72. See also Peter Sloterdijk, *Essai d'intoxication volontaire* (Paris: Hachette, 2006), 160.
29. Vinciane Despret has, however, considered how profoundly animals, groups of animals, and animal-human social interactions can change over time, such that animals have a history (*Quand le loup habitera avec l'agneau* [Paris: Les Empêcheurs de la penser en rond/Seuil, 2002]).
30. For Darwinians, animals change by genetic mutations, but animality remains what it is.
31. Alison Abbot, "Regulations Proposed for Animal-Human Chimeras," *Nature* 475 (2011): 438.
32. Dominique Lestel, *Paroles de singes* (Paris: La Découverte, 1995). Haraway's observation that we must be cyborg is accurate but falls a little short today; we must consider the multiplication not only of cyborgs, but of their assemblages with one another, and understand that we have always been hybrids, but differently at different times.
33. To my knowledge, it was Sherry Turkle who first drew attention in 2006 to this strange phenomenon.
34. On the question of "Life in Common," see Dominique Lestel and Hollis Taylor, eds., *Social Science Information* 52, no. 2 (2013): 183–6.

# Animality and Contemporary French Literary Studies

## *Overview and Perspectives*

ANNE SIMON

*Translated by* CÉLINE MAILLARD AND STEPHANIE POSTHUMUS

Whether reflecting on human animality and the interactions between humans and animals in literature, questioning the potential for creative language to express non-human affects and relationships to the world, examining the reconfigurations of anthropocentrism, or even contemplating "the end of human exceptionalism,"[1] collective research on animality in literature in France has flourished since the mid-2000s. The scope of this research is unprecedented, as evidenced by the "Animots" research project that will be discussed later. The novelty of this research is not only its focus on the animal question, which has long been conspicuously absent from literary criticism, but also its methodology. It was initially based on an interdisciplinary approach that led to the formation of new corpuses, the reconsideration of the animal in literature of the last century, and the establishment of new cross-disciplinary research on different forms of animal knowledge (relationships between creative writing on animals and history, ethology, law, ethics, paleontology, and more typically philosophy, to name a few). In this burgeoning field, problems are addressed collectively and internationally, with numerous contributions by North American and more generally Anglo-Saxon researchers, as well as by networks supported by French academic bodies that often include researchers affiliated with foreign institutions.[2]

The purpose of this article is to report on such growth; it is time to offer an overview of completed research projects and their preferred approaches, as well as of emerging perspectives in twentieth- and twenty-first-century French literary studies. These chronological boundaries may seem narrow; however, they will allow me to better define the currency of the animal question in France, as illustrated by the dramatic increase of art exhibitions, seminars, lectures, conferences, since the beginning of the twenty-first century.[3] Given the state of literary studies on animality in France today, offering a comprehensive view of the topic from the Middle Ages to the present day would be impossible, as this research is still largely emerging. In other words, the present article can only serve as an introduction, with no claim to exhaustivity.

It is a positive and significant sign that in recent years, a small number of academic networks dedicated to the study of animality in literature have been created, some of which

*Le roman rustique animalier* : **relations entre humains et animaux dans les récits ruraux de langue française (XXe et XXIe siècles)**

**Colloque international**

**Du 14 au 16 novembre 2012**

**9h30 - 16h30**

Organisé par **Alain Romestaing** et **Alain Schaffner**, « Écritures de la modernité » (EA 4400 - Université Sorbonne Nouvelle - Paris 3 / CNRS).

Dans le cadre du Programme « Animots : animaux et animalité dans la littérature de langue française (XXe et XXIe siècles) » soutenu par l'Agence Nationale de la Recherche (ANR).

**Mercredi 14 et jeudi 15 novembre 2012 - Salle Extérieure**
**Maison de la Recherche de Paris 3**
*4 rue des Irlandais - 75005 Paris - Métro (ligne 7) : Place Monge - RER (ligne B) : Luxembourg*

**Vendredi 16 novembre 2012 - Salle Las Vergnas (3e étage)**
**Université Sorbonne Nouvelle - Paris 3 - Centre Censier**
*13 rue de Santeuil - 75005 Paris - Métro (ligne 7) : Censier Daubenton*

***Entrée libre dans la limite des places disponibles***

*Figure 1.* Poster for the conference "Le roman rustique animalier" (The Animal Rustic Novel), part of the research cluster *Animots*, hosted by the Université Sorbonne Nouvelle in Paris, November 2012, organized by Alain Romestaing and Alain Schaffner. (Source: Alain Romestaing, used with permission)

have received financial support, and thus symbolic legitimacy, at various institutions. Two projects have been specifically devoted to animality in modern and contemporary literature: "Animalittérature," the first international research project in this area in France, which resulted in a second project, "ANIMOTS: Animals and Animality in 20th and 21st Century French and Francophone Literature," funded by the French National Research Agency.[4] Other scholarly initiatives have taken up broader questions in a diverse array of disciplines, some of which include literature, notably the series of seminars entitled "Frontiers of Humanity/Frontiers of Animality,"[5] which focused on animals in nineteenth-century France, and "Human-Animal Interactions,"[6] which emphasized dance and the arts. The research project "Animal Flesh: Approaches to Animality in Art and Literature"[7] includes literature as well, but without focusing on French texts. Many more research clusters and conferences could be named.[8]

The vitality of the animal question in art and literature stems from a cooperative effort that includes all academic disciplines. The American and British historians Peter Sahlins and Chris Pearson evoke, not surprisingly, the broader notion of "French animal studies" even as they work more specifically in historical studies.[9] In the present article, however, I am most interested in outlining the current state of literary studies of the twentieth and twenty-first centuries in terms of their institutional legitimacy in France.[10]

## POSSIBLE PATHS FOR COMPARING ANGLOPHONE AND FRENCH DIFFERENCES

Animal studies, human-animal studies, critical animal studies: research programs deriving explicitly from these approaches, which cover a field of thought much broader than mere literary studies, still remain rare in North American and Anglo-Saxon universities. However, the fact that these areas can be listed individually indicates that over the last decade, the animal question in the English-speaking world has begun to gain academic recognition, in large part because it combines well with earlier areas of study, such as ecocriticism and cultural studies. In France as well, many humanities and social sciences (HSS)[11] researchers have been engaging, for several decades, in a lively discussion about animality. In literary studies, a much smaller number of researchers have been working individually on the animal question since the late nineties, and it has only been since the mid-2000s that this research has been institutionalized. It is a telling fact that it was not until 2011 that the need arose to find a French translation for the expression "animal studies" as part of the "Animots" research project. (In passing, no French translation exists for the expression "human-animal studies."[12]) Translating the expression "animal studies" became the object of collective reflection during the seminar "The Animal between Sciences and Literature,"[13] and yet no unanimous decision has been reached. This indicates, if not a discomfort—what is more exciting than becoming aware of a major intellectual if not political issue!—then at least a difficulty in finding a definition for animal studies that would be immediately relevant and meaningful to the French. There were those who proposed no translation, either because they were affiliated with Anglo-Saxon or North American institutions, or because their work in the field made them doubt the need for a cultural transposition, or because "Études animales" referred to veterinary science. And then there were those who proposed various translations—"Études animales," "Études animalières," "Études sur l'animalité," "Étude de l'animal"—while at the same time delimiting the boundaries of these possibilities. Finally, it is the expression "Études animales" that has won a small majority so far. I concur with this choice, insofar as the term "Études" (studies), rarely used in France to designate an academic discipline, keeps track of a cultural transfer and highlights the reality of a global dialogue. The adjective "animales" has the paradoxical advantage of being less specific than a noun complement, and so it can express the fact that animality (including human animality), animals, or human/animal relations will be studied using various methodological and critical approaches. This type of adjectivization is not perfect in French, but has a precedent found in the now-accepted expression "Éthique animale" (animal ethics), in the (less common) "Philosophie animale" (animal philosophy), and in "Sciences humaines et sociales" (humanities and social sciences). I believe it is crucial to use a translation that emphasizes differences in terms of methods, corpuses, and culture with respect to the English-speaking world—especially North America—not for the purpose of creating a rivalry between the two, but in order to transform these differences into a source of inspiration for reflection (nothing is more productive for thinking than cultural relativizing). Moreover, it is important for researchers working on an emerging topic to have an "official" affiliation, for practical reasons (financial and logistical support) but more significantly for symbolic ones (visibility for a research topic within the context of national research policy). Thus, the aim was not so much to adopt the practices of Anglophone thinking into French research, as to draw on them in order to form conclusions and confront issues that were similar, not so similar, different, divergent, etc.

Does the fact that this discussion about an acceptable translation for animal studies occurred so recently in France indicate that there is some kind of delay in terms of the legitimacy of the animal question? To the extent that many disciplines have been confronting the animal question for a long time in France, the latter answer is no (I will return to this); however, in literary studies, a *reluctance* that is heavy with meaning can be detected when it comes to including animality. This reluctance is similar to the one in the area of gender studies; it has only been in the last two or three years that the field has been translated (as "Études de genre") and acquired some academic recognition.[14] Even if, as I noted earlier, the term "studies" acts as a flag of North American origin, the use of the word "genre" within the French institution, long dominated by literary critics who opted for an "écriture féminine," indicates an evolution towards an understanding of the social construction of gender. The reluctance to translate gender studies can be explained then in terms of an initial ideological stance that, once transformed, "authorized" the translation process.

The fact that the translation of animal studies is only now becoming an issue is different, however, from the case of gender studies. First, we should note that animal studies in the Anglophone world are not first and foremost connected to literary studies (whereas this is the case for gender studies). Literary scholars then did not necessarily look to literature for approaches and inspiration. Instead, researchers in French literary animal studies have been interested in ecocriticism and environmental studies in the last few years (I'm thinking here of colleagues working in the areas of *géopoétique*, Anglophone studies, and comparative literature in France, especially those working on Anglo-Saxon and American literature from an ecocritical perspective).[15] All things considered, there are many reasons that could explain literary scholars' reluctance towards animal studies and the question of animality, a question that is often considered unworthy of being brought into a discussion about the power of language, or reductively opposed to concerns about the human.

I would like to continue for now with the comparison with North American and Anglo-Saxon intellectual communities. A first reason for such a reluctance might be that in the French tradition, literary studies do not establish a theme or a political cause as a separate academic field, as is the case in the North American world (I'm thinking here of gender studies, queer studies, Jewish studies, to name a few examples). There are, however, groups of scholars in France (including myself) whose work has been deeply influenced by these areas of study. Coming back to gender studies, Audrey Lasserre suggests that this area may have attained the status of a separate discipline in France because of the indirect support of European research policy and research grants. The resistance to "studies" may be due to the French universalist tradition that fears, rightly or wrongly so, the reduction of a group to one single specificity and the essentialization that may result when ignoring the intersections that make a group more of a questioning node than a stable ontological state; researchers therefore prefer to not separate these questions from other issues. In short, research in France has been defined primarily by its disciplines (philosophy, literature, history), its methods, or more recently, its disciplinary intersections.

At the same time, research in France has tended, from the start of its reflections on animality, to connect this question to the environmental question, whereas in the English-speaking world, animal studies and ecocriticism have developed separately. It should, however, be noted that in France there was no "first wave" of ecocriticism,[16] a development rooted in North American thinking and whose militant, activist approach reflected a 1970s political environmentalism, with an emphasis on the axiological value of literary works that would be negatively characterized in France as "romans à thèse." Other discrepancies that should be noted are: early ecocriticism's return to a simplified notion of referentiality and the transparency of literary language after the North American infatuation with postmodernism; the focus on nature writing, which has no real equivalent in the French literary tradition and which is unfortunately relatively unknown in France except by American literary scholars; the mythification of wilderness, which has been revealed to be more cultural myth than was first thought, and which in any case has no real or conceptual equivalent in France (I'm thinking here of the very different histories of *réserves naturelles* and national parks in the two countries, and of the significant distinction between the French term *paysage* or "countryside" and the North American "nature").[17] As Alain Suberchicot notes, "There are . . . countless books about bears, wolves, hunting dogs, and caribou herds in American environmental literature,"[18] and, I might add, about humans living in harmony with a wild, natural environment. Early ecocriticism, very rooted in North American culture, has not had any influence on the field of research on animality in France. Indeed, its restrictive point of view would not have been of much use when analyzing, for example, the representation of stray dogs in suburban garbage dumps or in conflict zones in Jean Rolin's work, the girl/cat hybridity in Béatrix Beck's *L'Enfant-Chat*, or the dead boy's metamorphosis into a terrifying and polymorphic sea animal in Marie Darrieussecq's *Bref séjour chez les vivants.* The entrenched opposition between supporters of humanism and proponents of biocentrism, as well as the opposition between ecocentrism and biocentrism continue to be the subject of North American debate. In France, on the other hand, humanism is not seen as a synonym of anthropocentrism, nor is it opposed to a theoretically valid approach to the living world more generally. While many French researchers, including myself, work critically on anthropocentrism, the question is not raised in the same way in France, where sociobiology and evolutionary psychology remain suspect when applied to literature.[19] Similarly, controversies around the concept and idea of "nature" that have recently arisen in the Anglophone field of ecocriticism can be linked to a debate about hyper-constructivism born from North American postmodernism and French theory. Even if the adjective "French" may suggest otherwise, these two movements were institutionalized in the United States and did not develop in the same way in France.[20] (The English expression "French theory" is used in France, without a translation.) French scholars have been extremely wary of the artificial regrouping of extremely diverse thinkers, and the simplification of what are in fact very different issues. The supposed opposition between "deconstruction" or "relativism" on the one hand and some "biological given" on the other is not relevant for analyzing the question of animality in Derrida or Deleuze and Guattari.[21] And even if the *diktat* of the textual has hidden those critical movements that have been attending to the question of animality, there is a tendency today in France to reread the thinkers of the sixties and seventies without rejecting outright their theoretical and methodological contributions. The development of a literary history with new methods, objectives, and a renewed attention to the world *hors texte*, represents a major shift that does not, however, entail the total rejection of structuralist work whose analysis of form and genre remains fundamental to the study of texts, including those on animality.

Despite these divergences, it is now possible to speak of cross-pollination between these two disciplines; researchers working in France on animality are now not only interested in animal studies, but also in ecocriticism's "second wave"[22] and its commitment to expanded boundaries.[23] By including works that are less directly about natural spaces[24] (and that may even give priority to urban or abandoned spaces), by rethinking the definition of nature, by examining the relationship between the global and the local and between social and environmental injustice, in short, by nuancing the overly rigid opposition between humanism and environmentalism, ecocriticism in its present-day form is much more likely to interest French literary specialists who wish to analyze, for example, Yves Bichet's *La part animale*,[25] a novel narrating the hazards of intensive turkey farming, or Antoine Volodine's stories, in which animals' and humans' interchangeable bodies share a bleak post-apocalyptic world.

## ISSUES SPECIFIC TO THE FRENCH RESEARCH CONTEXT

Before highlighting the research projects that are now opening up, I will outline some issues that are specific to the French context and that may help to explain literary studies' reticence towards the animal question. French literary critics' focus on self-referentiality and textual language games since the sixties and seventies first marginalized the importance of thematic criticism with a phenomenological interest in the relation of humans to the "perceptible world"[26] or to animality (even their own). Seen as the apex of human language, the highest form of figurativeness, literature did not have much to do with animals, who were seen as too "silent" or too resistant to human language to interest formalism. Mosquitoes, chickens, sheep, and even the noble lion seemed unworthy subjects of study for literary criticism that systematically reduced animal stories to a form of children's literature, or that has long focused on human drama or great political narratives. It should be noted that the animals named above play a significant role in the works of authors as important as Albert Cohen (who raises issues of God's creation, the animal soul, and the right to kill), Marcel Proust (who raises questions about guilt and sadism), Jean Giono (who touches on violence and war), and Joseph Kessel (who explores human omnipotence and human/animal eroticism).

In contrast to this portrayal, I underline as I did earlier that an important number of disciplines in the humanities and social sciences have been rethinking the animal question for several decades, some that were originally and constitutively marked by this issue (such as philosophy, anthropology, or paleontology), and others that have historically had less to do with it (such as sociology or psychology). Indeed, the animal question, as well as the question of human/animal relationships, has seen, since the early nineties, rapid development in terms of academic publications and legitimacy, as illustrated in the work of historian Michel Pastoureau, sociologist Dominique Guillo, philosopher Florence Burgat, anthropologists Frédéric Keck and Noélie Vialles,[27] anthropologist Philippe Descola (who received a gold medal from the National Center for Scientific Research [CNRS] for his work on Western issues "beyond nature and culture"),[28] to name but a few. In philosophy in particular, recurring and sometimes heated debates[29] oppose supporters of human specificity, like Étienne Bimbenet, who would keep humans above and away from animality, and advocates of a continuist approach like Georges Chapouthier. Rather than taking sides in this perhaps oversimplified dispute, I would like to note that literary criticism has been rather silent, whereas writers, from Colette to Valère Novarina, Louis Pergaud, or Jean-Loup Trassard, have never been silent.

I should point out that some scholars have identified the importance of animality in twentieth-century literary works. But their studies have tended to systematically reduce the animal to the human by analyzing the animal "character" as allegory or symbol (a deer as human desire, a female cat as jealousy), or by focusing on the peasantry or the hunting world within a regionalist approach, or by limiting their perspective to certain genres considered secondary (children's tales, fables, bestiaries, the "rustic novel," the "animal story," natural-history writing). The conventional nature of these approaches did not allow for epistemological shifts, or reconsideration of the literary and historical corpus, which are essential to contemporary literary studies on animality.[30] The latter consider diverse interdisciplinary perspectives essential, some foundational as in the case of phenomenology, which has long served as a source of key references, and some more original as in the case of the conversation that is beginning to happen with disciplines such as the life sciences, field ethology, and cognitive sciences.

On a thematic level, there has been particular interest in the interactions and sharing of meaning between humans and animals; the motifs of monstrosity and hybridity (both of which allow the reader to experience other corporealities and psyches);[31] the question of innate animality (Darwinist, phenomenological, psychological); and the representations and self-portraits of writers as animals, from Proust (owl, hen, or wasp) to Eric Chevillard (hedgehog), from Marie Darrieussecq (sow) to Tristan Garcia (monkey). On a formal level, there has been particular interest in the ways in which some authors forgo anthropocentrism, trying to put into words animal subjectivities or consciousness,[32] or trying to represent animal languages (breaths, rhythms, significant movements) and environments,[33] or trying to report on animal "actions" and "styles."[34] These attempts give rise to a series of questions: Can human language, taken to its highest degree of complexity as the vehicle of literary innovation, give voice, body, and affects to other species? Can projection beyond self and empathy, both seen as unique to the creative process, provide access to a specific otherness without falling into anthropomorphism and its supposed flaws, an idea that many scientists and thinkers are now rethinking, and rightly so?[35] Finally, on a social and political level, there has been particular interest in the concept of biopower as defined by Michel Foucault and revisited by Gilles Deleuze: this perspective examines relationships between *genre*/gender and animality or animalization, between dehumanization and the denaturing of animals, but also between the performativity of speech and the capacity of writing that refuses pathos, in the work of such writers as Maryline Desbiolles, Yves Bichet, Marie Darrieussecq, Patrick Modiano, Marie-Hélène Lafon, Jean Echenoz, and Olivia Rosenthal.[36] It should be noted in particular that literary scholars can now raise the issue of violence to animals, not only in the wake of contemporary writers who care about animals, but also in the wake of great humanist thinkers and writers who initiated the debate in the nineteenth century (from Jules Michelet to Émile Zola, but also including Victor Schoelcher and Victor Hugo), and who were followed by others in the twentieth century (Colette, Pierre Gascar, Marguerite Yourcenar, Romain Gary). This relatively slow development contrasts with the academic legitimacy that has been given to the animal question in the Anglophone world, where animal protection and welfare have followed a different trajectory. Researchers need to analyze not only the stylistic devices that bring animal suffering to our awareness, but also, historically, the ideology behind the erasing of the issue in the formation of the twentieth-century literary canon as presented in textbooks and classrooms. Finally, literary studies on animality in France are influenced by contemporary philosophy's recentering on emotion, ethics, and the issue of "care" (which is still struggling to find academic legitimacy, but will certainly do so in the coming years).

Literature's intersection with history, neglected during the years of formalism, is being reassessed in order to determine the impact of the animal question on writers. Scholars have observed a renewed interest in traditional genres in the work of contemporary authors, such as Pierrette Fleutiaux, Alain Mabanckou, Patrick Chamoiseau, Jacques Roubaud, Jacques Lacarrière, Jean-Pierre Otte, Luc Lang, Tristan Garcia, and Patrice Nganang, who have reclaimed narratives such as the tale, the bestiary, the satire, the fictional manual of zoology, and the Ovidian metamorphosis model. A critical review of the literary canon of authors considered "worthy" of study is also underway:[37] revisiting the work of writers such as Jules Renard and Beatrix Beck, whose work was rarely studied at university; rereading major authors in whose work the animal question was minimized (Marcel Proust, Jean-Paul Sartre, Albert Cohen);[38] and reconsidering the connection between literary texts and the scientific and intellectual discoveries or sociopolitical events of their time (such as the impact of Darwinism, industrialization of rural land, extinction of species, intensive livestock production, scientific experiments, etc.).

The theoretical body of work being referenced and used to legitimize this research in France is different from the Anglo-Saxon and North American canons, although cross-cultural dialogues have thankfully been bringing about change. So far in France, the work of thinkers who do not separate philosophical thought from poetic creation has served as an inexhaustible source of reflection for literary scholars. To name a few examples, I will mention Maurice Merleau-Ponty's draft of *The Visible and the Invisible* and his seminar "La Nature";[39] Jacques Derrida's posthumous book *The Animal That Therefore I Am*[40] (which we no doubt know too well for its neologism "animot"); Carlo Ginzburg's famous article on the indicial paradigm that touches on anthropology's story about the relationship between hunting and storytelling in the process of hominization;[41] Élisabeth de Fontenay's *Le silence des bêtes*,[42] and Jean-Christophe Bailly's *Le versant animal*.[43] These thinkers' descriptions of the multiple ways in which humans build a common world with animals (note the importance of Husserl's notion of the "arch"), their insistence on the animal's elusion in its relationship to the world and to humans, their emphasis on the need to consider animals in their uniqueness, and their denunciation of different forms of violence affecting not only animals but also indirectly humanity's way of seeing itself, have proven to be important underlying elements of literature about animals since the beginning of the twentieth century. These intersections of poetry and philosophy are at the heart of the concept of *oikos*, represented more metaphorically than realistically by French writers, and largely influenced by Germanic thought. I'm thinking here of the crucial importance of nature in German poetry and philosophy since the late eighteenth century, and of the role of the latter in the education of many French intellectuals. For example, Hölderlin, for whom poetic language is the only way to really "inhabit" the Earth; Hofmannsthal and Rilke, both authors whose poetry expresses the perilous human relationship with animal modes of being, even if these latter turn out to be too empathically invested or on the contrary dramatically inaccessible, are all "traditional" referents for French scholars. Thoreau, Emerson, Whitman, Leopold, on the other hand, represent only a limited source of inspiration for a small number of French writers and thinkers.

Finally, I should note that there are other thinkers who are regularly cited by literary scholars, but who are more directly related to disciplines like natural history, life sciences, and ethology: most frequently, Charles Darwin, Jakob von Uexküll, and Dominique Lestel; less often but with equal significance, Michelet, Fabre, Maeterlinck, Konrad Lorenz, Adolf Portmann; and in a more philosophical vein, Hans Jonas and Vinciane Despret. Authors such as Gaston Bachelard, Bruno Latour, and Michel Serres, all French writers, are taken up more often in the Anglophone world than in France, where they receive too little attention as far as I'm concerned. Such an observation reaffirms the need for international collaborative research programs, which are key to weighing and considering these different theoretical frameworks. Perhaps with undue optimism, I also hope for publishing policies that recognize this need for more translations of important texts in each cultural context. In any case, a work in both English and French on Francophone literary texts (while awaiting other possible combinations, such as German and French) is probably the only way to create real overlaps between different cultural fields; this is one of the objectives of the research project "Animots" that I am currently directing and that has publications available in both English and French—this is particularly true for the 20th–21st Century French and Francophone Studies International Colloquium "Human-Animal/Humain-Animal" (figure 2), which was attended by nearly three hundred scholars mostly from literary disciplines, and which resulted in three collective works in both English and French.[44]

Because the topic being addressed is *literary* animality, French scholars situate their reflections less in the general framework of ecocriticism, and more in the specific context of ecopoetics, or even biopoetics and zoopoetics. (Time will tell if these terms, not yet used often and as of yet not precisely defined, will prove to be empty shells or not.) These scholars examine, then, the capacity of literary language to represent the strength of what connects us to animals as well as to represent the extraordinarily diverse ways animals inhabit the world. As such, they are particularly interested in the ways in which literature combines with other forms of existing knowledge (natural history, ethology, ethics, biology, politics of nature), but also creates, through embodied stories, a strictly *literary knowledge* of these animals as "flesh and bones, claws and fur, smells and cries," to cite the first pages of Élisabeth de Fontenay's *Le silence des bêtes*,[45] as "singular animals" whose individuality has been dissolved by the sciences in general and the humanities and social sciences in particular through their use of the practical but too "generic"[46] and abstract concept of "animality."

**20th/21st Century French and Francophone Studies International Colloquium**

Wednesday, March 30 - Saturday, April 2, 2011
The Westin St. Francis Hotel, Union Square, San Francisco

# HUMAN-ANIMAL / HUMAIN-ANIMAL

*Cheval No. 2* © J. Tomas Lopez

University of San Francisco
Centre national de la recherche scientifique
Agence nationale de la recherche

*Figure 2.* Front page of the program for the 20th–21st Century French and Francophone Studies International Colloquium, organized by Anne Mairesse and Anne Simon in San Francisco, 2011. (Source: Anne Simon, J. Tomas Lopez (artist), used with permission)

NOTES

I would like to thank Louisa Mackenzie, Stephanie Posthumus, Miranda Mueller, and Éliane DalMolin for their thorough proofreading of this article.

Some terms including the word *studies* were in English in the original French text, signaling the lack of any direct equivalent in French-speaking context of these academic areas.

1. Jean-Marie Schaeffer, *La fin de l'exception humaine* (Paris: Gallimard, 2007).
2. For example, the "Animots" research project includes eight researchers of whom three are from Anglophone institutions in the U.S. and the UK (Princeton, University of Connecticut, and Roehampton University).
3. See for example the conference "Que la bête meure!" (The beast must die), organized by Philippe Dagen and Marion Duquerroy, at the Institut National d'Histoire de l'Art/Musée de la Chasse et de la Nature, Paris, June 11–12, 2012. Thanks to its curator Claude d'Anthenaise, The Musée de la Chasse et de la Nature is becoming a larger public venue for animal studies in art and literature.
4. "Animalittérature" (2007–2010), led by Anne Simon, Sorbonne University, Nouvelle-Paris 3, and "Animots: Animaux et animalité dans la littérature de langue française (XXe–XXIe siècles)" (2010–2014), led by Anne Simon. This international research project is affiliated with the École des Hautes Études en Sciences Sociales (EHESS), and in partnership with the Sorbonne University, Nouvelle-Paris 3.
5. "Frontières de l'humanité/frontières de l'animalité," organized by Claude Millet and Paule Petitier at the Université Denis Diderot-Paris 7 from 2007 to 2012, which led to an online publication: *L'animal du XIXe siècle* (The Nineteenth-century Animal), ed. Paule Petitier, http://www.equipe19.univ-paris-diderot.fr/Colloque%20animal/Page%20titre%20colloque .htm.
6. "Animal/humain: Passages," organized by Daniele Meaux and Jean- Pierre Mourey at the University of Saint- Etienne, 2010–2012; selected papers published in *Figures de l'Art* 27 (2014).
7. "La chair de l'animal: Approches de l'animalité dans l'art et la littérature," led by Valérie Boudier and Gille Froger within the Tourcoing "Pôle Arts Plastiques," 2012–2013.
8. For example, the conference "The Sense of the Animal" (*Le sens de l'animal*), held at the University of Poitiers in 2010 with proceedings published as *La question animale: Entre science, littérature et philosophie*, ed. Jean-Paul Engélibert, Lucie Campos, Catherine Coquio, and Georges Chapouthier (Rennes: Presses Universitaires de Rennes, 2011); the conference "The Time of the Living" (*Le moment du vivant*), organized by Arnaud François and Frédéric Worms at Cerisy-La-Salle in August 2012; the conference series "Learning Saturdays—Thinking the Living" (*Les samedis des savoirs—Penser le vivant*), organized by Roland Schaer at the Bibliothèque Nationale de France in October 2012; the research projects "Literature and Knowledge of the Living—19th-20th Centuries" (*Littérature et savoirs du vivant—XIXe-XXe siècles*), led by Gisèle Séginger at Université Paris Est/Fondation Maison des Sciences de l'Homme; and "Thinking the Living: Exchanges between Literature and Life Sciences from the Late 18th Century to the Present Day" (*Penser le vivant: Les échanges entre littérature et sciences de la vie de la fin du XVIIIe siècle à l'époque contemporaine*), led by Gisèle Séginger and Christine Maillard, Fondation Maison des Sciences de l'Homme. I would like to thank Christine Baron for the information she provided.

9. See "Animals in French History," *French History*, vol. 28, no. 2 (2014).
10. I will not mention the work of individual researchers since they have remained outside a larger global approach to animal literary studies.
11. "SHS" (Sciences humaines et sociales) have no strict equivalent in the Anglo-Saxon world (we will refer to them simply as the "humanities" and "social sciences"). SHS includes a large number of disciplines (literary theory, philosophy, history, anthropology, sociology, law, musicology, art history, epistemology, linguistics, political science, geography, archeology, paleontology, psychology, economics, etc.) and one institute at the CNRS (Centre National de la Recherche Scientifique, or National Center for Scientific Research) is specifically devoted to research in this area.
12. The theoretical differences of human-animal studies and animal studies will not be discussed here.
13. "L'animal entre sciences et littérature," a seminar given as part of the "Animots" research project, led by Anne Simon at the EHESS in 2011–2012. The work will be continued in the master's program seminar "Animal History: Animality and French Literature (20th-21st Centuries)" (*Histoires de bêtes: Animalité et littérature de langue française [XXe-XXIe siècles]*) at the EHESS, 2012–2014.
14. Only since 2011 has the CNRS offered positions in section 35 (*Sciences philosophiques et philologiques, sciences de l'art*) entitled "Gender Studies" (*Études de genre*). Moreover, when Audrey Lasserre and I published the compendium *Nomadismes des romancières contemporaines de langue française* (Paris: Presses de la Sorbonne nouvelle, 2008), and even earlier when I published with sociologist Christine Détrez *À leur corps défendant: Les femmes à l'épreuve du nouvel ordre moral* (Paris: Seuil, 2006), we still adopted the English expression "gender studies."
15. See the work of Alain Suberchicot, Yves-Charles Grandjeat, Michel Granger, and Jean-Paul Engélibert.
16. Lawrence Buell, *Writing for an Endangered World: Literature, Culture, and Environment in the U.S. and Beyond* (Cambridge, MA: Belknap, 2001), 17. According to the author, the two ecocritical "waves" are not strictly successive; "movements" may be more appropriate. On the relationship between ecocriticism's origins and 1970s environmental activism in the USA, and evolving convergences with French ecological thought, see Stephanie Posthumus, "Translating Ecocriticism: Dialoguing with Michel Serres," *Reconstruction: Studies in Contemporary Culture* 7 (2007), http://reconstruction.eserver.org/072/posthumus.shtml.
17. See François Specq, "Henry D. Thoreau et la naissance de l'idée de parc national," *Écologie & politique* 36 (2008): 29–40; and Pierre Schoentjes, "Textes de la nature et nature du texte," *Poétique* 164 (2010): 477–479.
18. Alain Suberchicot, *Littérature et environnement: Pour une écocritique comparée* (Paris: Champion, 2012), 118.
19. For a comparative study of the two cultural contexts, see Kerry H. Whiteside, *Divided Nature: French Contributions to Political Ecology* (Cambridge, MA: MIT Press, 2002), who clarifies in particular the French ecological view as affiliated with "noncentered ecologism" or "ecological humanism"; Stephanie Posthumus, *La nature et l'écologie chez Lévi-Strauss, Tournier, Serres* (PhD diss., University of Western Ontario, 2003), http://mcgill.academia.edu/StephaniePosthumus/Papers/510412/La_nature_et_lecologie_chez_Levi-Strauss_Tournier_Serres; Lucile Desblache, "Penser et représenter les animaux: Cultures anglophones et francophones," in *La plume des bêtes: Les animaux dans le roman* (Paris: L'Harmattan, 2011), 85–98.
20. See François Cusset, *French Theory: Foucault, Derrida, Deleuze & Cie et les mutations de la vie intellectuelle aux États-Unis* (Paris: La Découverte, 2003).
21. Jacques Derrida, *The Animal That Therefore I Am*, trans. Marie-Louise Mallet (New York:

Fordham University Press, 2008); Gilles Deleuze and Félix Guattari, *A Thousand Plateaus: Capitalism and Schizophrenia*, trans. Brian Massumi (Minneapolis: University of Minnesota Press, 1987).

22. Buell, *Writing for an Endangered World*, 17.
23. Karla Armbruster and Kathleen Wallace, eds., *Beyond Nature Writing: Expanding the Boundaries of Ecocriticism* (Charlottesville: University Press of Virginia, 2001).
24. British philosopher Kate Soper relativizes the concept of nature in *What Is Nature? Culture, Politics, and the Non-Human* (Oxford: Blackwell, 1998).
25. Yves Bichet, *La part animale* (Paris: Gallimard, 1994).
26. Jean-Pierre Richard, *Proust et le monde sensible* (Paris: Seuil, 1974).
27. Frédéric Keck and Noélie Vialles, eds., *Des hommes malades des animaux* (Paris: L'Herne, 2012).
28. Philippe Descola, *Par-delà nature et culture* (Paris: Gallimard, 2005).
29. Schaeffer, *La fin de l'exception*, and Catherine Larrère and Raphaël Larrère, *Du bon usage de la nature: Pour une philosophie de l'environnement* (Paris: Aubier, 1997), which have raised much debate around this controversial issue.
30. For example, the conference "Le roman rustique animalier" in Paris, November 2012 (see figure 1).
31. See *Hybrides et monstres: Transgressions et promesses des cultures contemporaines*, ed. Lucile Desblache (Dijon: Presses Universitaires de Dijon, 2012).
32. One of the subjects of the conference "Le roman rustique animalier."
33. Jakob von Uexküll, *Milieu animal et milieu humain*, trans. Charles Martin-Fréville (Paris: Rivages, 2010).
34. See Jean-Christophe Bailly, "Les animaux conjuguent les verbes en silence," in *Facing Animals/ Face aux bêtes*, ed. Anne Mairesse and Anne Simon, special volume of *L'Esprit créateur* 51 (2011): 106–114; and Marielle Macé, "Styles animaux," ibid., 97–105.
35. See Brian L. Keeley, "Anthropomorphism, Primatomorphism, Mammalomorphism: Understanding Cross-species Comparisons," *Biology and Philosophy* 19 (2004): 521–540; Françoise Armengaud, "L'anthropomorphisme: Vraie question ou faux débat?" in *Les animaux d'élevage ont-ils droit au bien-être?* ed. Florence Burgat and Robert Dantzer (Paris: I.N.R.A., 2001), 203–231; Jacques Dewitte, "L'anthropomorphisme, voie d'accès privilégiée au vivant: L'apport de Hans Jonas," *Revue philosophique de Louvain* 100 (2002): 437–465.
36. See Anne Simon, "Hommes et bêtes à vif: Trouble dans la domestication et littérature contemporaine," in *Le moment du vivant*, ed. Arnaud François and Frédéric Worms (Paris: Presses Universitaires de France, 2013).
37. See for example Lucile Desblache, *Bestiaire du roman contemporain d'expression française* (Clermont-Ferrand: Presses de l'Université Blaise Pascal, 2002); and Jacques Poirier, ed., *L'animal littéraire: Des animaux et des mots* (Dijon: Presses Universitaires de Dijon, 2010).
38. *Cahiers Albert Cohen* 18 (2009), "Animal et animalité dans l'œuvre d'Albert Cohen."
39. Maurice Merleau-Ponty, *Le visible et l'invisible* (Paris: Gallimard, 1964); and Merleau-Ponty, "La Nature: Notes, Cours du Collège de France," ed. Dominique Séglard (Paris: Seuil, 1995).
40. Derrida, *L'animal.*
41. Carlo Ginzburg, "Signes, traces, pistes: Racines d'un paradigme de l'indice," *Le Débat* 6 (1980): 3–44.
42. Élisabeth de Fontenay, *Le silence des bêtes: La philosophie à l'épreuve de l'animalité* (Paris: Fayard, 1999).
43. Jean-Christophe Bailly, *Le versant animal* (Paris: Bayard, 2007). Questions raised by the book were addressed during the panel "Imagine a world without animals?" organized by Éliane Dal-Molin at the MLA annual convention in Boston in January 2013.

44. "20th–21st Century French and Francophone Studies International Colloquium," organized by Anne Mairesse and Anne Simon in San Francisco, March 30–April, 2011. The publications are *Contemporary French and Francophone Studies (Sites)* 16 (2012): *Human/Animal Part 1*, ed. Roger Célestin, Éliane DalMolin, and Anne Mairesse; and *Human/Animal Part 2*, ed. Roger Célestin, Éliane DalMolin, and Anne Simon; and the special volume of *L'esprit créateur* 51, no. 4 (2011), "Facing Animals/Face aux bêtes."
45. De Fontenay, *Le silence des bêtes.*
46. Jean-Christophe Bailly, *Le dépaysement: Voyages en France* (Paris: Seuil, 2011), 359.

# PART 3

# Animal Intimacies

# Why "I Had Not Read Derrida"

## *Often Too Close, Always Too Far Away*

VINCIANE DESPRET

*Translated by* GRETA D'AMICO AND STEPHANIE POSTHUMUS

There are, wrote Jacques Derrida in 1997, "two grand forms of theoretical or philosophical treatise regarding the animal . . . In the first place there are texts signed by people who have no doubt seen, observed, analyzed, reflected on the animal, but who have never been *seen seen* by the animal. Their gaze has never intersected with that of an animal directed at them (forget about their being naked) . . . That category of discourse, texts, and signatories (those who have never been seen seen by an animal that addressed them) is by far the one that occurs most abundantly. It is probably what brings together *all* philosophers and all theoreticians *as such*." Derrida continues: "As for the other category of discourse, found among those signatories who are first and foremost poets or prophets, . . . who admit to taking upon themselves the address that an animal addresses to them, . . . I as yet know of no statutory representative."[1]

The critique that Donna Haraway directs at Derrida, some ten years later in the book *When Species Meet*, renders legible the fact that the discourses designated by Derrida are not on the order of two, but rather three.[2] In fact, because he is discussing the possibility of taking seriously the act of being seen by an animal and the consequences of this when writing about the animal, Derrida groups together on the one hand scientists *and* philosophers, and on the other, prophets and poets. Haraway questions, "Why did Derrida not ask, even in principle, if a Gregory Bateson or Jane Goodall or Mark Bekoff or Barbara Smuts or many others have met the gaze of living, diverse animals and, in response, undone and redone themselves and their sciences?" Two lines later, she provides in the form of another question the answer implicit in the first: "Why did Derrida leave unexamined the practices of communication outside the writing technologies he did know how to talk about?"[3]

In so doing, Haraway very clearly undoes that which constitutes the first category: she separates philosophers from theoreticians, first, and then from scientists. She thus renders perceptible an essential difference. The first group will always speak in the absence of the animal, while the second face to face with the animal. Of course, and Haraway underscores this point, speaking "face to face" offers no guarantee: one can work *on* animals without seeing their gaze, but one can also work *with them*—and Haraway provides some examples of this, taking seriously the fact that the animal looks at us, and that this look should be returned.

By complicating these categories, by asking Derrida to distinguish philosophers from scientists and to distinguish between types of scientists, Haraway reveals clearly Derrida's original positioning, a positioning that he himself explicitly evokes: he proposes to speak (starting) from [*à partir de*] a *real animal* and not about *animality* (being animal), to speak in the animal's presence and not in its absence. This positioning gives rise to a thunderclap in the clear skies of the philosophical tradition relating to the animal.

Derrida speaks about the animal against his tradition. This is not, in itself, particularly original; it is part of the very game of philosophy, to fight with ancestors and contemporaries about great and noble ideas, the question of human exceptionalism and radical otherness having the potential to place among the ranks of these noble ideas. But to do as Derrida did is far more original and risky; it is impossible to not be sensitive to this situation when one has been, as I have, embedded in this tradition.

The topic of the animal in the French tradition is considered philosophical only under certain conditions, which I am conscious of simplifying (but this simplification itself gives a fairly precise idea of what is and is not acceptable); the animal can be a topic of philosophy if it is called up as a figure of otherness (or deprivation) or as our ancient ancestor, the image of a nature whose traces we would have retained, but in this case, the search for difference is unavoidable, without appeal. The animal may thereby enter philosophical discourse under the condition that it not be evoked for itself, but rather as a figure caught up in theoretical and abstract issues. It can achieve credibility under the condition of "keeping up appearances" in the most literal sense.

Derrida's taking up of the topic of the animal in order to oppose a certain humanist hegemony and denounce inequality and violence was, then, in the French philosophical universe, not only original but also daring and risky.[4] All denunciations and appeals for other forms of consideration or care concerning the topic of the animal put the author at risk of being relegated to the dungeons of moral philosophy—second-rate philosophy considered much less noble than true conceptual work—or worse still, of being accused of giving in to sentimental or feminine tendencies, whose political commitment can only be inherently suspect.[5] By denouncing philosophical violence towards the animal, Derrida distinguished himself clearly from this tradition.

But the true originality of Derrida, which I mentioned when discussing Haraway's recategorization, resides in his rejection of the philosophical game *par excellence*, most explicitly the one that deals only with representations. Derrida insists on this in the very first pages of his book: he does not speak of animals as generalities; the cat of whom he speaks is not a *figure* that can be applied to all cats on earth, nor is it a matter of felineness. He speaks even less of the "animality" that traverses our religions, myths, literature, and fables. Derrida hopes to close the parenthesis on "the history of the philosophical animal,"[6] which deals not with real, but abstract, inexistent animals: he calls these "animots," which is to say "paper" animals set in motion by modern Western philosophy in order to think hierarchically, not to think and experiment with others. To make this shift, Derrida calls for replacing the instrumental approach with an engagement in questions genuinely concerned with animals themselves.[7] In short, Derrida proposes nothing less than a break with the strictly philosophical—or epistemological—regime of "representations."

When this book was published, I had not read Derrida. And I did not read him in the years that followed. And yet I was a philosopher who was engaged in philosophical work—in any case I was paid to do philosophy—and my work concerned animals.

## THE TWELFTH CAMEL

For over ten years, I did not *want* to read Derrida. Not until 2009 did I finally acquire his book. When I say "I did not want," I do not mean to say that I was unaware of his work—how would this have been possible?—in fact, older colleagues had recommended it to me time and again. Nor did I miss the scarcely concealed reproach behind this recommendation: Wasn't it time I did *real* philosophy? I admit that this only increased my suspicion. If these colleagues, who were embedded in the tradition I just described, thought that it would be useful for me to read Derrida, it followed that little could come from reading him.

It was not until the moving experience that came about by reading Haraway's *When Species Meet* that I finally decided to show interest in the matter. I will not speak much of Derrida in the present article—others are far more qualified to do so—but I will refer to him in light of my odd refusal and in order to ask this simple question: What happened to me? By asking this question, and trying to respond to it, I will take stock of the manner in which I have fit my work, more or less so, faithfully at times, with resistance to others, in this philosophical tradition that is mine.

Even if I will not talk about Derrida, I will explain the reason for the quotation at the beginning of this article. With a clarity that still astounds me, Derrida outlines the reasons that I held dear when categorically refusing to read the philosophers of my tradition who spoke about the animal. Given my refusal, which, paradoxically, Derrida's text acknowledges as well-founded, I could not have known that I would find an ally by reading him. I will start with what is most surprising: it would take not only more than ten years but also the intercession of an American author, who does not write in my language and who does not belong to the tradition in which I work, to force me to come back to authors from my tradition, my language, and my field of research. It would take, in other words, a long journey elsewhere before I could agree to return home. In the present article, I will try and understand why.

A parable told to me by Isabelle Stengers over fifteen years ago has since embedded itself in the heart of my work. It is the parable of the twelfth camel.[8] I believe that this parable sheds light on the strange detour I had to make in order to come back to those thinkers from my own tradition, that is, French philosophy. I should add that this story will not elucidate the full meaning of my journey; rather it will serve as a guide, outlining the threads and different stages.

> An old man, sensing his impending death, called to his side his three sons, to share with them the little he still owned. He said to them: my sons, I have eleven camels. I bequeath half of them to the oldest, a quarter to the second son, and to you, my youngest, I give a sixth. Upon the father's death, the sons found themselves quite perplexed: how to divide their inheritance? A war over the division of goods seemed inevitable. With no apparent solution, the sons went to a neighboring village to seek advice from an old sage. The old man thought awhile and then shook his head: I cannot resolve this problem. The only thing I can do for you is to give you my old camel. He is not very obedient and often does as he pleases. I don't know if you can use him but I think he may help you divide your inheritance. The sons brought the old camel back with them and divided up the inheritance: the first then received six camels, the second three and the youngest two. This then left the camel of the old sage, which they could return to his owner.

This parable reveals a very particular and essential dimension of all forms of inheritance: they place us in a position of obligation. And this is exactly what the sons work out: they are obligated[9] by their father's inheritance, and all the more so because the inheritance appears impossible, unless it is destroyed or rejected altogether. What their experience illustrates is that this inheritance passed on as such requires them to "start from"[10] this inheritance. The preposition "starting from" has different implications than those of other prepositions. "Starting from" does not mean the same thing as "about," or "concerning," or "with." It implies precisely the fact of remaining obligated to that *from* which we speak, think, or act. It implies letting ourselves learn from events and creating from them. Learning from events and being obligated to them means first and foremost learning to do, speak, act, and decide, not *about* these events, not *concerning* them, not *facing* or *against* them; it is to learn to do, speak, act *from* them, in an act of creation. We are bound. The sons were *bound*. And this explains precisely why they resisted the temptation of common sense—as well as far less common-sense solutions, such as rejecting the inheritance, cutting the camels into pieces, or killing one another—why they resisted the temptation to find a compromise for the problem, why they honored the terms of the problem such as it was posed along with its contradictions. They understood that what their father had passed on to them was built upon the fact that inheriting is an act that demands thought and commitment, an act that calls for our transformation by the very gift of inheriting.

This parable is connected to my own story in two ways. First, by simple analogy: indeed, I traveled far, searching for what I needed in order to accept an inheritance from Derrida, which in fact my own tradition had bequeathed to me. As for the terms of this inheritance, they will not be discussed here because I have done this work elsewhere, putting to the test what I had read of Derrida through Haraway, in particular by studying the way in which scientists were beginning to respond to their animals *and* becoming attentive to the animals' responses in turn.[11] What Haraway says about the way Derrida reads the tenet "thou shalt not kill," about the way he "so passionately read and reread so that it could not be read the same again,"[12] I could say about her reading of the philosopher's text, which I inherited. To inherit is indeed an act that demands a transformation on the part of the inheritor (see for example, in figures 1 and 2, the illustrations of a seven-year-old friend upon hearing the parable).

But this fable can just as well impart another meaning to my story. This meaning will not come into play until much later in the present article; I simply mention it here in passing. The twelfth camel appeared in such a way as to put an end to the compromise I had made between two contradictory obligations: that of thinking *from* philosophy, and that of thinking *from* animals.

However, before exploring this connection between the parable and my story, I would like to highlight what could have been a third connection. This more obscure point shows that the act of being bound through inheritance that I am describing also contains many contradictions. The camels that are the object of the inheritance are, as it were, excluded from the problem. If I insist on the fact that inheriting takes place "starting from," it is nevertheless not *from* them as camels per se, living animals with whom these sons will learn to live, that this story is constructed. It is constructed around them, about them, perhaps even "with them," due to the fact that cutting them into pieces would seem an absurd decision (but if these were books, the solution would be equally absurd, which shows that I have not fully considered the living specificity of the camels). In my taking up the story, I am not referring to any actual engagement to the camels; these latter will not be transformed by the inheritance. They will simply be handed down. To think "from them" would require that, for example, the dignity and engagement entailed by the act of inheritance changes something for the camels,

*Figure 1.* The parable of the twelfth camel, retold by Amaya, age seven. The story is about the importance of transforming that which is transmitted to us, that which we inherit. In this spirit, we told the story to our young friend. Without prompting, she too transformed what she had been given and drew her new version of the story. In this first drawing, as she explained, "The third son should be given the twelfth camel as a gift, so things can be more fair. And when I started to draw him, I realized I was drawing myself, so the son is actually a girl. And the old camel turns out to be a mother who had a baby. So now the girl can have a camel and a baby, and can help take care of them both." We were so impressed by this nascent desire for equality and care that we decided to illustrate the chapter with Amaya's transformations of the parable.

*Figure 2.* THE PARABLE OF THE TWELFTH CAMEL, RETOLD. THIS DRAWING "SHOWS EVERYONE. THE OLDEST SON HAS SIX CAMELS. THE MIDDLE SON HAS SPIKY HAIR BECAUSE HE WANTED TO BE A BIT DIFFERENT TOO. THE YOUNGEST DAUGHTER IS HOLDING THE GIFT CAMEL AND HER BABY CAMEL."

guarantees them worthy and steady trainers in the sons, who by learning to respond responsibly to their father, learn at the same time to respond responsibly to their animals. I accepted, without critique, the terms under which the problem was put forward, and in particular the fact of speaking of "owners." I am not innocent in this respect, and this makes me an heir to a tradition against which I nevertheless try very hard to fight. In my earlier reading, it is quite possible that the camels are in fact "animots," and that to inherit from this story requires us to take it up again and again, to show the limits and contradictions, as, in fact, our young illustrator has done (in figures 1 and 2).[13]

## THINKING *FROM* ANIMALS

If there is something that philosophy cannot or does not want to do, and Derrida underscores this, it is indeed to think *from* animals, in the sense that I intend this preposition to be understood. When philosophers profess to speak from animals, the meaning of "(starting) from" changes; it no longer signifies the same thing at all, and the double meaning of the French term (*à partir de*) comes fully into play: it is to *depart from* animals, to leave them as quickly as possible.[14] And certainly never to return. The animal becomes text and pretext, that from which we speak and that which we immediately forget. Its only function is to provide a reason for going (*partir*) elsewhere.

I myself have "started from" the animal in this way, and as my analysis of the fable shows, I am not yet entirely over this way of doing things. But if I did things this way for so long, it is because it seemed to me a condition of doing philosophy, and not a condition of the other disciplines such as science or literature. The recurring accusation from other philosophers that "this is not philosophy" still holds much weight.

So I continued to do things as usual and in relative comfort. But at a certain moment in my journey, I was no longer able to. The animals began to oblige me, for reasons that I will explain shortly. I could no longer follow the oppressive philosophical custom of speaking of "animality," the abstract animal, the absent animal, the symbolic animal. I could not *not* break away from the framework of "representation" within which the animal was discussed in philosophy. From that point on, if I felt obliged to account seriously for the presence of these animals, and thus to learn to speak of real animals, and if I wanted to do this *from* philosophy, I could not do so except under certain conditions—and so emerged a compromise. And of the conditions of this compromise, one stood out more than others: I had to do this speaking of real animals while maintaining a distance.

Maintaining distance was a prudent measure, justified and legitimate, at the time. This caution is a characteristic of the episteme of the French philosophical tradition: getting involved is highly suspect, as is showing any sort of sentimentality or sensitivity, which is destined to win you the crown of martyrdom. I was not particularly interested. Even less so, I confess—and it is only in responding to the invitation to write this article that I fully realized this—because I had internalized several of the norms that regulate our manner of writing and thinking, even feeling a certain disdain for those who risked being marginalized.

Here is where the parable of the camels comes back in: the twelfth camel appeared in such a way as to put an end to this compromise. Responding to two contradictory demands, *starting from* the animal and *starting from* philosophy, I fell into the trap that the sons had successfully avoided. I had tried to divvy up the inheritance according to an arrangement unworthy of the problem; this arrangement found a pragmatic solution instead of taking into account the possible dimensions of the problem, what it required and what it involved. The compromise did not honor philosophy, because it made those who practice it amputate part of themselves, nor did it honor the experiences that interested me, i.e., those of impassioned scientists who followed the lead of their animals and who took the risk of responding to their animals' needs.

## ENCOUNTERING A BIRD

For the author-turned-ethologist, one of the classic genres of ethological literature written for a general audience is the story of how it all started, whether this be a childhood passion for animals, or a deeply moving encounter with an animal that revealed a new destiny, as in Saint Paul's encounter with God on the road to Damascus. I fear that my article may seem to share many similarities with this genre—road to Damascus, take two. And in fact, it will take on a certain resemblance at various moments, a point I concede entirely. However, it is not a question of me complying with the rules of a genre of literature with which I have long been familiar. It is a matter of performing through narration the pressing obligation that is now mine: to always attempt, by all means possible, not only *not* to erase the presence of the animal, but above all to avoid relegating the animal to the status of a passive object. This is a moral, political, and epistemological obligation, the terms of which I have learned from Bruno Latour and Isabelle Stengers. It has also evolved, more recently, into an aesthetic obligation. I have become an amateur, in the Latourian sense: a person who likes and cultivates her tastes and does her best to cultivate a kind of becoming-sensitive to the world.

I have learned that in ethology, and more generally in animal sciences, monologues make terrible narratives. I nevertheless produced many of my own monologues, when I began to work as a philosopher on the "animal question." This expression, as suspect as it is for me today, seemed at the time the most viable way of bringing the animal into philosophy. The title of the first article I published in English indicates clearly the philosophical tradition in which I was situated: "Ecology and Ideology: The Case of Ethology."[15] I was searching for traces of ideological and political contamination in the work of scientific naturalists. It goes without saying that I did not have any intention of speaking about the real animal. Rather, my aim was to follow the path of representations, the symbolic, and to track down all that threatened the separation of science and politics. I had not read Latour. I was resolutely on the side of epistemology as it had been taught to me. I had discovered the projects of Patrick Tort, who had put together a group of researchers analyzing the work of Darwinian naturalists (such as the anarchist Kropotkin, a perfect target for this sort of critique).[16] When culture (or politics) enters nature, the epistemologist is duty-bound to show it the door.

From a contemporary perspective, I was able to bring this research to bear on ethology. Debates on sociobiology were raging. It has since become clear that these critiques adopted different positions depending on their country of origin. Stephen Jay Gould and others devoted themselves to pointing out the scientific poverty of sociobiological models, while in France, opponents tackled the political implications and, in particular, the perceived threat to "human dignity" due to the human/non-human animal connection being made in an admittedly cursory way by sociobiologists.[17]

The fact remains that it was with the intention of "separating the wheat from the chaff" that I went to the Negev Desert in Israel to work with an ethologist, Amotz Zahavi, who was studying a bird that appeared to be rather unusual, the babbler. My objective was simple: to do in the field what philosophers did with respect to texts. I did not intend this to be a significant change in my research, except inasmuch as it would lend an anthropological dimension to the study, which I hoped would be original. And I would certainly take more pleasure in observing scientists in beautiful desert landscapes than in reading about them in libraries.

I had good reasons to think that Zahavi would be the perfect victim for the sort of critical approach that I had learned to carry out theoretically. He credited his birds with unusual abilities. First, babblers are altruists and cooperate in many different types of situations. Of course, sociobiological literature had prepared us to accept this about birds;[18] babblers, however, do this in a remarkably more inventive and diversified way and for entirely different reasons than "sociobiologized" birds. They do it for reasons of prestige. Other birds, much more sensibly, content themselves with evolutionary reasons—they do it to pass on their genes, according to the theory of kin selection. Babblers thus deviate from the norm that rules bird behavior.

Second, the extravagant behavior of these birds and their ethologist did not stop there: babblers danced, Zahavi claimed. And they danced together, dangerously showing off in plain view of their predators, because this was their means of testing the strength of their bonds, and of building these bonds. Such hypotheses were bold when discussing baboons;[19] they seemed highly improbable when discussing birds. Besides, to claim that birds *danced* could only reveal the existence of a mechanism such as "projection"—such an explanation fit perfectly with my critical emphasis on "representation." To see birds as "dancing," and for fairly complex reasons, could only have been a result of the fact that the observers projected onto the animals their own frameworks and experiences. My critical analysis would prove to be that much easier because Zahavi made himself so available to such criticism: his anthropomorphism was everywhere, without restraint (he credited birds with the most complex intentions, and complex intentions always seem human), and he did not hesitate to make comparisons between the situation of these birds and the origins of kibbutzim.

So I left to join Zahavi and his babblers. And I saw the birds dance.

## SEEING BIRDS DANCE

As Isabelle Stengers so beautifully put it: "Being in the field happened to me/came upon me." As my work with Zahavi progressed, the birds acquired more and more importance. It was not the ethologist that I needed to observe; it was the birds *and* their ethologist. Together. For it was together that they were fascinating. The birds made Zahavi interesting, and Zahavi had been worthy of the interest of the birds. He had made them fascinating. From the narrative he proposed on their behalf, more hypotheses were created, more questions arose, more beings became necessary to successfully describe the situation. This is what the work on texts had entirely erased: the living presence of those about whom, or rather, *from whom* scientists speak, and by whom they are authorized to speak. I discovered that any theory of "representations" was at once partial and totalizing, because it proposed to elucidate the complex work of relations and encounters from the sole standpoint of the human.[20]

How could I continue to do philosophy starting from this very different perspective? Fortunately, I discovered Isabelle Stenger's *The Invention of Modern Science*, and thanks to her, the work of Bruno Latour.[21] I needed to become interested in actual practices, in the way they articulated questions and responded to these questions. Of course, I understand that the stories scientists develop about animals are also our stories; but these stories transform humans and their animals. They are embedded within a specific temporality, attuned as sciences of contemporaneity, explains Isabelle Stengers, as sciences for which the production of knowledge is also a production of a way of being. They do not reveal what animals are; they follow and accompany an act of becoming together, an act of becoming with the stories that we construct concerning them. These stories of transformation must agree grammatically with the future anterior: birds *will have been* far more interesting starting from the moment that Zahavi proposed to connect their stories to others; sheep *will have* been far more sophisticated starting from the moment that Thelma Rowell asked them interesting questions. And these will not be simple theoretical questions that at best modify the way we perceive animals (although this matters from the outset, and particularly for farm animals). To ask interesting questions, it is necessary to create conditions in which sheep are able to demonstrate an interest in these questions. Do they have friends, as we have said of chimpanzees and baboons? Give them the time to organize themselves, put together a flock that will make it possible, give them the space and freedom to organize themselves at will, do not make them compete over food.[22] Interesting research looks at the conditions that allow beings to become interesting.

At this point, if I wanted to fully understand the transformations these animals underwent when under the scrutiny of our practices, I had to follow closely the way scientists spoke to animals, how they made animals agents, how scientists created the conditions for certain responses with respect to what was being asked of the animals. In other words, I needed to accept the requirement of trying to understand how these changing animals became real by way of the very test of transformation that had been proposed to them; to understand the system of truth that was at the heart of these tests; to understand how these animals were involved in what William James calls the process of verification.[23] Otherwise, I would simply be producing an umpteenth critical analysis of "representations."

I learned to have confidence in scientists. Not all scientists, of course, but in those who, as Bruno Latour said about the work of Thelma Rowell with her sheep, "give the animals a chance."[24] And I learned to be interested in what it was that gave rise to this increased interest, this transformation.

Accepting such a requirement meant that I was under the same constraints as those in whom I had placed my confidence. When asked to explain why her work with baboons has produced such interesting results, Shirley Strum answers that, first of all, she made an effort not to construct knowledge about the baboons "behind their backs"; in her practice, questions asked of the baboons were subordinate to the need to know "what mattered to them."[25] This courtesy of "getting to know what matters" has proven successful, and so I in turn am compelled to try it: baboons become interesting when scientists work under this constraint; I could in turn hope, in my analysis, to make the researcher interesting by adopting the same constraint, and by exploring how "what matters to them" allowed the transformations to occur. And of all the things that "matter" to scientists, there was one that I could not ignore: how animals take an active part in the knowledge that is produced about them. By taking an interest in the activity of the researchers, I thereby became interested in the activity of their animals.

## THINKING WITH TWELVE CAMELS

Even if practicing courtesy met my requirement of doing philosophy without mutilating or weakening those I wished to account for, it nevertheless still followed the essential rule of distance. I can now say that "being in the field happened to me/came upon me." I could not have said this a few years ago, no more than I would have taken the risk of revealing the upheaval I felt upon seeing those inventive and remarkable birds, or the happiness of sheep, cows, and pigs,[26] or the sadness of captive wolves in a park in the Lorraine, not because they were captive but because captivity had transformed them into stupid and cruel beings. I can now say what I could not have said a few years ago, even if describing this sadness and knowing the judgment it will bring is not without contradiction or uneasiness. I can now accept and understand what animal technicians are telling me when they say they love their animals and take care of them. And I can also say of certain scientists that not only do they do "bad science" ["*mauvaise science*"]—which remains a way of keeping distance—but that they do "science bad(ly)" ["*science mauvaise*"].

I cannot say that I love animals, any more than I can say that I love children. But I have loved many people who were children. And I have loved many creatures that happened to be animals.

So I began to transgress boundaries. Something about today's intellectual climate no doubt allows me to do this more than I could have in the past. I also believe that the influence of Anglo-Saxon animal studies—and I am thinking here, for example, of the work of Cary Wolfe and many others who should be mentioned as well, who were not as cautious, who agreed to commit themselves to closing the distance—has favorably influenced this shift in Francophone circles. Reading Donna Haraway—the story of her love for her dog Cayenne, what this love requires of her, the questions it raises and the thoughts it generates—was the type of event that rarely occurs in a lifetime. Learning to think *from* love. Touching and thinking *from* the fact of having been touched. Discovering grace through thinking and without innocence. It could be done. And it could be done *from* philosophy, without dishonoring it. One could do it while continuing to think. Or perhaps even while being all the more obligated.

It was not a matter of coming out of the closet. There is no closet. There are transformations. No conversions, but perhaps a toppling over, falling from a horse, on a path that no longer leads to Damascus. Saul does not answer God; he politely asks him to wait two minutes while he comforts his horse. Let God wait; other things matter. This is where true conversion takes place. This would be the true miracle of "becoming idiot," as Gilles Deleuze puts it.

A horse or a camel—on the condition that, in the case of the latter, I rethink the fable of inheritance and consider seriously how the camels themselves were able to obligate the sons—each of them creates an opening, opens up a possibility, asks the question "and if?" that undoes previous threads and reconnects them to something else. And the problem takes on an altogether different meaning, obliges you to do things differently, to do things otherwise. This is not a "pretty story," it is simply a story that can be given a sequel, a story that obliges the person who hears it to also then be in a position of obligation.[27] The twelfth camel puts an end to my compromise that was meant to resolve a set of contradictions; it reopens these contradictions and leaves me with the question: "What do I do with this?" The twelfth camel is not lacking a sense of humor. He proposes an experiment, in which you have no way of knowing whether you will be the actor or the subject.

In order to explain the effects of this twelfth camel as clearly and concisely as possible, I will conclude by coming back to the last lines of an article I devoted to Haraway's book, and that was published in a journal of French philosophy. The fact that I wrote this article, drawing on my own narrative thread, and that it was accepted in an academic review, reveals the possibility of other philosophical futures, with animals, other ways of thinking, and above all, another geometry, more variable, less restricted, more inventive, and certainly one of obligation rather than distance.

"It is towards a becoming-sensitive that Donna Haraway works. Thinking with more living beings, sensitizing ourselves to other creatures, finding other ways of living in the world. I have proposed a Spinozist reading of Haraway's writing technique—a remedy for indifference and contempt—because this writing and the paths it invites us to take seem to suggest, create, or give rise to an extension of the possibilities of becoming attached to the multiple threads that make up the fabric of the world, to an increase in the possibilities of the powers of existence, of others and our own. One does not emerge unscathed from reading this kind of book. A blackbird sang beneath my window every morning this past spring. Reading Haraway obligated me to hear it. For this blackbird sang as if the world itself depended on its song, and the importance of things came to dwell in its voice. This blackbird, who conversed with others in the joy of dawn, made *importance* exist in another way: importance became incorporated in the world, this last spring. Undoubtedly it had been there for a long time, but I needed an encounter in order to be pierced by it. And this 'importance' rose like a question: how can I now write in such a way as to be worthy of what matters, with a similar insistence, for another being? How can I in turn bring into existence, in a transformative way, that which was able to touch me? Life invades philosophy; life asserts its importance in, imports itself into, philosophy. And the fact that I dare evoke this type of experience illustrates the effects of what demands to be continued: the feeling of becoming a passageway for new connections."[28]

## NOTES

1. Jacques Derrida, *The Animal That Therefore I Am*, trans. David Wills (New York: Fordham, 2008), 13–14.
2. Donna Haraway, *When Species Meet* (Minneapolis: University of Minnesota Press, 2008), 21.
3. Haraway, *When Species Meet*, 21.
4. It is true, as Stephanie Posthumus points out, that the works of Claude Lévi-Strauss had already initiated this critique, and that anthropology began the very important work of decentering the figure of the white occidental male (and of critiquing ethnocentrism). But this work has been limited to the domain of human relations, where this critique appeared naturally legitimate (even if controversial). It was much less the case when the animal also became an issue.
5. The incredible difficulty that American feminist thinking has had achieving recognition within French and Belgian academic circles reveals, I think, a similar sort of reticence.
6. Derrida, *L'animal que donc je suis* (Paris: Galilée, 2006), 42.
7. It is not within the scope of my article to revisit Haraway's critique that Derrida does not see his project through to the end, and that he lacks curiosity about his cat. I defer to Haraway on this point.
8. This parable first guided my doctoral work on emotions, a part of which has been published in English and which contains the parable analyzed from a considerably different perspective; see Vinciane Despret, *Our Emotional Makeup: Ethnopsychology and Selfhood*, trans. Marjolijn de Jager (New York: Other Press, 2004), 1–36. Isabelle Stengers has also included the parable in her book *Penser avec Whitehead* (Paris: Seuil, 2002). Isabelle and I have subsequently worked from the perspective of this parable in a book that we wrote together: *Les faiseuses d'histoires: Que font les femmes à la pensée?* (Paris: Les empêcheurs de penser en rond/La Découverte, 2011), translated to English by April Knutson: *Women Who Make a Fuss. The Unfaithful Daughters of Virginia Woolf* (Minneapolis: University of Minnesota Press: 2014).
9. The term "obliged/obligated" brings to mind the distinction that Isabelle Stengers makes between the idea of *obligations* and that of *requirements*. All activities have *requirements* with respect to the world in which they are practiced in order to exist. For example, experimental sciences require a laboratory in order to produce a "reliable witness" and to create a difference between a scientific object and an artifact. But not all activities arise from *obligations*, which is to say they do not all put themselves at risk, nor do they all make an effort to present themselves politely. Being obligated means agreeing to expose oneself to failure, refusing to construct the words of order that would protect oneself from the requirements of the activity. Stengers underscores that this "distinction [between obligation and requirement] is a thing to *be made*, and not acknowledged as already manifest in the state of things" (Isabelle Stengers, *La vierge et le neutrino: Les scientifiques dans la*

*tourmente* [Paris: Les Empêcheurs de penser en rond, 2006], 230). See also Isabelle Stengers, *Cosmopolitiques 1: La guerre des sciences* (Paris: La Découverte, 2003), 60. The system of obligations cannot simply be transferred over to the system of animal rights; even if we can say that animal rights subject us to obligations, these rights do not obligate us in the Stengersian sense. This is because an obligation is always a thing to be done, to be created, to be invented, and no response can, nor should it hope to, conclude the debate, hesitations, and actions. An obligation is always in this sense more demanding than a right. Concerning this topic, see Vinciane Despret and Serge Gutwith, "L'affaire Harry: Petite scientification," *Terrain* 52 (2009): 142–152.

10. Translator's note: *à partir de* in French.
11. Vinciane Despret, *Penser comme un rat* (Paris: Quae, 2010).
12. Haraway, *When Species Meet*, 20.
13. I thank Louisa Mackenzie for having stressed this point, which had been completely invisible to me. This blindness reveals the difficulty of the problem of this inheritance I am receiving and reconstructing, and in fact attests to the long road yet ahead if we are to truly think "starting from."
14. Translator's note: the author articulates the double meaning of the term as follows: "c'est à partir des animaux, *à partir au plus vite*."
15. Vinciane Despret, "Ecology and Ideology: The Case of Ethology," *International Problems* 63 (1994): 45–61.
16. Patrick Tort, ed., *Darwinisme et société* (Paris: PUF, 1992).
17. Of course, this summary contrasts only too quickly the broadest tendencies. American feminists, for example, critiqued the patriarchal, conservative, and masculinist ideology that underpinned sociobiological discourse. But they avoided abstract declarations about the "dignity" of humans and were aware of their own agenda. Concerning this topic, for primatology, see Donna Haraway, *Primate Visions: Gender, Race, and Nature in the World of Modern Science* (New York: Routledge, 1989).
18. I should clarify that I knew ornithological literature well because of the dissertation in ethology that I had written for my degree in psychology.
19. Barbara Smuts and John Watanabe, "Social Relationship and Ritualized Greetings in Adult Male Baboons (*papio cynocephalus Anubis*)," *International Journal of Primatology* 11 (1990): 147–172.
20. Bruno Latour's second principle of symmetry will allow me to formulate this point attentively. The bibliography of the book I wrote shortly after my return from being in the field clearly reveals my orientation then: 92 of the 124 titles were written by scientists working with animals, and most of them dated back at most ten years. Vinciane Despret, *Naissance d'une théorie éthologique: La danse du cratérope écaillé* (Paris: Les empêcheurs de penser en rond, 1996).
21. Isabelle Stengers, *L'invention des sciences modernes* (Paris: La Découverte, 1993); Bruno Latour and Steve Woolgar, *Laboratory Life: The Social Construction of Scientific Facts* (Beverly Hills, CA: Sage Publications, 1979); and Bruno Latour, *Nous n'avons jamais été modernes: Essai d'anthropologie symétrique* (Paris: La Découverte, 1991).
22. See Vinciane Despret, "Culture and Gender Do Not Dissolve into How Scientists Read the World," in *Rebels, Mavericks, and Heretics in Biology*, ed. Oren Harman and Michael Dietrich (New Haven: Yale University Press, 2008), 338–355; and "Sheep Do Have Opinions," in *Making Things Public: Atmospheres of Democracy*, ed. Bruno Latour and Peter Weibel (Cambridge: MIT Press, 2006), 360–370.
23. William James, "Pragmatism's Conception of Truth," *Pragmatism: A New Name for Some Old Ways of Thinking* (London: Longman, 1907), 197–236.

24. Bruno Latour, "A Well-Articulated Primatology: Reflections of a Fellow Traveler," in *Primate Encounters: Models of Science, Gender, and Society*, ed. Shirley Strum and Linda Fedigan (Chicago: University of Chicago Press, 2000), 358–382.
25. For an overview of changing attitudes in primate research, see Shirley Strum and Linda Fedigan, "Changing Views of Primate Society: A Situated North American View," in Strum and Fedigan, eds., *Primate Encounters*, 3–49.
26. Vinciane Despret and Jocelyne Porcher, *Être bête* (Arles: Actes Sud, 2007). This book was written following a study we conducted with breeders and their animals.
27. Once again, I reveal the fact that I am an inheritor of a philosophical tradition that tends to not refer to law (or to not expect solutions from law because it creates other considerable problems). The system of law is undoubtedly remarkably inventive, as Bruno Latour argues, and I certainly celebrate the emergence of more intelligent and sensitive forms of legislation that do a better job of protecting animals. But the risk of this inventiveness is that it may lead, in real terms, to less inventiveness necessitated by the fact of being obligated (and certainly by the fact that certain species will be considered worthy of rights and others not). I am in this respect inspired by the work of Emilie Hache, who draws on William James to define pragmatism as an *art of consequences* that is interested in the *effects* that its propositions induce in order to verify veracity. In terms of law and in light of Hache's pragmatism, I will practice the skill of the casuist: it is a matter of carefully describing moral situations *in existence* or *in the process of becoming*. It is not a matter of saying or stipulating what should be done, but trying to carefully describe *what people are doing*, becoming witnesses for those who are caring and bringing about change, and knowing that this testimony participates in the process of change. See Emilie Hache, *Ce à quoi nous tenons* (Paris: La Découverte, 2011).
28. Vinciane Despret, "Rencontrer, avec Donna Haraway, un animal," *Critique* 747–8 (2009): 747–759.

# *Chercher la chatte*
## *Derrida's Queer Feminine Animality*

CARLA FRECCERO

This essay situates some of the dilemmas of the effort to think with non-human animate being in the Western philosophical tradition by examining the posthumous work of Jacques Derrida, *The Animal That Therefore I Am.*[1] I argue for the usefulness of Derrida's work on animality for crafting a queer ethics of relating to the living in general, just as his notion of spectrality offered a way to grapple with the traumatic persistence of (historical) affect in the present. Nevertheless, even as Derrida reaches toward a referent by insisting on the particularity and singularity of *his* (female) cat, what he animates is the lively density of intertextual feline figures in the history of literary and philosophical thinking and writing about questions of figure and reference and questions of inscription (as "cat scratch"). Thus, even as Derrida seeks to literalize the allegorical search for the elusive figure of the animal other as a mode of *chercher la femme*, his work subtly demonstrates the figural inter-implications that shadow the discourses of feline femininity (involving both sex and species difference) in efforts to meet and face animate alterity.

Derrida's theoretical practice can be understood to be "always already" queer theory, if queer theory is understood in one of its valences, that is, as an immaterial de-normativization that works at the level of language, thought, and ideology to critique, but in a viral fashion, by replicating terms and repurposing them so that their operation moves down paths that are overgrown with the bushes of normative philosophical thought.[2] These paths are inscriptions, they don't quite open up; but they leave—or are—traces, and can be followed, like the tracks that Derrida is following in *The Animal.* Key terms that have emerged from deconstructive gestures include queer, though "queer," I think, carries with it—as the wind does scent—faint but specific whiffs of sex/sexual identity/sexuality. Derrida thus helps us to understand how theory is always already queer, and to affirm this queerness further.

Derrida's later work that moves toward the non-human living also has the potential to invest another queer theoretical domain—let's call it animal theory—with a meditation on subjectivity that brings with it traditions of Western philosophizing on the human and non-human. Animal theory is a queer theory in this respect: that it displaces humanism, de-normativizes subjectivity, and turns us toward not difference but differences, one of the most emphatic of Derrida's lessons having been the impossibility of a reference to "the" animal in favor of singular, differential, abyssal relations.[3] Derrida's thinking on—about and with—animals, by displacing the humanist subject (in both anthropocentric and universalist senses), can thus assist the ethical aspirations of a queer theory devoted to refashioning that subject in order to address other models of being or becoming and the ethical relations among them.[4] Derrida's deconstruction of the Western carno-phallogocentric subject, especially when he writes about the living in general, queers ontology and creates queer ontologies, relations of desire among the living.[5] At the very least, his legacy for animal theory will have been to be the Western philosopher whose introduction of the non-human living into philosophical and ethical consideration allowed for a meeting of posthumanism with ethology, ecological activism, eco-feminism, and other disciplines queered by their attention to the trivial, beside-the-point, non-eventfulness of minor daily ordinary histories and stories that are nonetheless part of the afterlives of trauma. As Donna Haraway says of these sorts of histories, "I think we learn to be worldly from grappling with, rather than generalizing from, the ordinary."[6]

Animal lovers, animal-rights philosophers, ethologists, and others delight in Derrida's staged scene of the encounter with an animal that is the occasion, the starting point, for his exploration of human and non-human in *The Animal.* Although Derrida notes that he has had many animals running through his *oeuvre* (an astonishing number, actually, which he documents in this essay), this is the first appearance of what he refers to as "a real cat":

> I must immediately make it clear, the cat I am talking about is a real cat, truly, believe me, *a little cat.* It isn't the *figure* of a cat. It doesn't silently enter the bedroom as an allegory for all the cats on the earth . . . An animal looks at me. What should I think of this sentence? The cat that looks at me naked and that is *truly a little cat, this* cat I am talking about, which is also a female, isn't Montaigne's cat either, the one he nevertheless calls "my [pussy]cat" [*ma chatte*] in his *Apology for Raymond Sebond.*[7]

What can a reader possibly make of this claim here to be talking about a "real" cat, a cat who follows Derrida into his bathroom each morning, and/or who is there, across from him in his morning nakedness, observing him, "just to see," observing him and "not hesitating to concentrate its [his] vision" in the direction of his sex?[8] On the one hand, I hear Derrida responding to Barbara Smuts's somewhat impatient chastising of Elizabeth Costello in *The Lives of Animals* (where it is also, notably, a question of philosophers, poets, and non-human animals): "Why doesn't Elizabeth Costello mention her relations with her cats as an important source of her knowledge about, and attitudes toward, other animals? . . . Whatever her (or Coetzee's) reasons, the lack of reference to real-life relations with animals is a striking gap in the discourse on animal rights contained in Coetzee's text."[9] In *The Animal,* then, the philosopher rectifies this error—an error for which this same philosopher indicts the history of Western philosophy,[10] that of not having considered being seen by an animal, an actual animal.

At the same time, Derrida positions himself in the philosophical genealogy that is not one, as he has done before, by citing—and then pursuing a digression on—the early modern philosopher, anti-Cartesian *avant la lettre,* Michel de Montaigne, whom he credits with having importantly posed the question of an animal's response, rather than her reaction: "Montaigne recognizes in the animal more than a right to communication, to the sign, to language as sign (something Descartes will not deny), namely *a capacity to respond,*"[11] and who returns, again and again, in this essay about the animal one is. Montaigne mentions his cat, feminized (as Derrida's cat will be), and ventures perhaps an even more radical supposition concerning her: "When I play with my cat, who knows if I am not a pastime to her more than she is to me?"[12] Derrida does not stay with his cat, something that elicits a complaint from Haraway: "He did not fall into the trap of making the subaltern speak . . . Yet he did not seriously consider an alternative form of engagement either, one that risked knowing something more about cats and how to look back, perhaps even scientifically, biologically, and therefore also philosophically and also intimately."[13]

But, as Derrida demonstrates, and at the very moment when he is at his most "ordinary," the articulation of a relation to the other within the inhuman technology of language, however much it is a "grappling with"—and for Derrida it is most certainly a grappling with—cannot avoid a generalizing whose agent may very well not be human at all.[14] Derrida's cat, remember, is not any of the other "literary" or "figurative" cats, not E.T.A. Hoffmann's, Sarah Kofman's, Montaigne's, Charles Baudelaire's, Rainer Maria Rilke's, Martin Buber's, Alice's (Lewis Carroll's)—though his text might in fact be a reading of *Alice in Wonderland*, he notes coyly—not Jean de La Fontaine's or Johann Ludwig Tieck's, though all of these cats and more do, in fact, find their reference in Sarah Kofman's *Autobiogriffures*, and that title, "Auto-bio-cat-scratches," if nothing else, might alert a reader to the sort of cat Derrida has in mind, if not before him.[15] The repeated negations Derrida pursues (maybe unconsciously, he playfully notes),[16] should alert a reader, make her suspicious—as the phenomenon of *dénégation* or negation must always do—that if this cat is not any of these other cats, if she is, "truly, believe me" a real cat, *not* the figure of a cat, then perhaps she is the cat of a figure. I will return to the question of what, exactly, she might figure as figure, later on in this essay. Derrida is also raising the question of whether it is ever possible to "represent" a "real" cat. In an essay where precisely the problem of the category, and categorization, of "the" animal is at stake, and where the singularity of the other is also at stake ("If I say 'it is a real cat' that sees me naked, it is in order to mark its [his] unsubstitutable singularity"[17]), what can he do but *write in* this conundrum of reference to the place where the question of referentiality must also be at stake?

The reference to the real also recalls a more famous cat, a cat who is recollected in the moments when Derrida writes, "for example," a cat ("for example the eyes of a cat," "for example a cat,"[18] etc.), the one on the mat that Alfred Tarski, J. L. Austin, John Searle, and a whole host of philosophers slung about as though they had it by the tail.[19] Given that this is an example used in philosophy and linguistics to talk about (semantic) truth and reference, an example of "positing" or "asserting," and that it features prominently in Austin's discussion of performatives and thus also recalls a famously conflictive moment in Derrida's own past (his debate with Searle in *Limited Inc*), it seems likely that Derrida's *dénégation* is just that—playful or otherwise.[20] Searle, it will be remembered, devoted an essay—"Literal Meaning"—to the sentence "the cat is on the mat," which came complete with line drawings of cats on mats in various poses and several time-space universes. Thus, even within what has often been referred to as Derrida's "ethical turn" and the effort, in this essay and the others devoted to non-human animals ("'Eating Well,'" "Violence against Animals") to meet animals in their phenomenality, to track them, as it were, and to respond to them as well, Derrida still and persistently returns to language and inscription, ironizing the effort to move beyond them even as he does so.[21] As he reminds his reader: "The letter counts, as does the *question* of the animal."[22] This is, in part, the reason why he also references the pseudonymous Lewis Carroll, logician Reverend Charles Lutwidge Dodgson, author of *Alice's Adventures in Wonderland* and *Through the Looking Glass*, whose Cheshire Cat and whose kitten also thwart Alice's efforts to pin down the "question" of the animal, language, response, and the real.[23]

And yet there *are* perhaps another few cats lurking in the background, and in particular a cat belonging to another dear other in Derrida's life whose articulations haunt his pages like the lost friend in Montaigne. Derrida's essay returns again and again to matters of living and dying, alterity, the absolute other, suffering, loss and mourning, to friends, neighbors, the proximate, in words that echo Derrida's writing on mourning and friendship across his many works:

> What is at stake in these questions? One doesn't need to be an expert to foresee that they involve thinking about what is meant by living, speaking, dying, being and world as in being-in-the-world or being-within-the-world, or being-with, being-before, being-behind, being-after, being and following, being followed or being following, there where *I am*, in one way or another, but unimpeachably, *near* what they call the animal.[24]

In the title essay of a collection that addresses, inter alia, the threats that seem to be posed by deconstruction—then most frequently referred to by the euphemism "theory"—to the U.S. academy, and that contains an interview where he distinguishes his work from that of Derrida, Paul de Man writes,

> It is a recurrent strategy of any anxiety to defuse what it considers threatening by magnification or minimization, by attributing to it claims to power of which it is bound to fall short. If a cat is called a tiger it can easily be dismissed as a paper tiger; the question remains however why one was so scared of the cat in the first place. The same tactic works in reverse: calling the cat a mouse and then deriding it for its pretense to be mighty. Rather than being drawn into this polemical whirlpool, it might be better to try to call the cat a cat and to document, however briefly, the contemporary version of the resistance to theory in this country.[25]

Here, in this other deconstructionist's text (and, as Wlad Godzich points out, deixis is also what is at issue), we find an amusing and late reference to the cat on the mat, in comic proximity to a paronomastic reference to Derrida as the mouse who is being derided.[26] In a further unmentioned reference to a discussion and disagreement between friends, this time Paul and Jacques rather than Searle and Derrida, de Man names Derrida as the man who, as in Rousseau, gets called a "giant" out of fear, a "tiger" who, rather than being a "paper tiger," is really a cat; there, in *Allegories of Reading*, it is also, as Godzich notes, a discussion about "the relation of figural language to denotation" (see figure 1). [27] This playful game of cat and mouse finds later echoes in *The Resistance to Theory*, but they are more ominous, even tortured, such as the overly stretched idiomatic phrase in the following: "Yet, with the critical cat now so far out of the bag that one can no longer ignore its existence, those who refuse the crime of theoretical ruthlessness can no longer hope to gain a good conscience. Neither, of course, can the theorists—but then, they never laid claim to it in the first place."[28] Or, more covertly, even unconsciously perhaps, in the stunning reference in the essay of the same name, "The Resistance to Theory," that renders undecidable the question of how papery this tiger actually is, not to mention whether the trap is for a tiger, a cat, or a mouse:

> Faced with the ineluctable necessity to come to a decision, no grammatical or logical analysis can help us out. Just as Keats had to break off his narrative, the reader has to break off his understanding at the very moment when he is most directly engaged and summoned by the text. One could hardly expect to find solace in this "fearful symmetry" between the author's and the reader's plight since, at this point, the symmetry is no longer a formal but an actual trap, and the question no longer "merely" theoretical.[29]

*Figure 1.* "EN AFRIQUE DU SUD, DANS LA RÉSERVE DE POTCHEFSTROOM, 1988." © DR. (SOURCE: CAHIER DERRIDA, L'HERNE, 2004, USED WITH PERMISSION)

De Man, like Derrida, also had a cat. And de Man, too, used cats as examples, usually ironic ones, as in these pages of *The Resistance to Theory*, a set of essays that documents the disturbance created by Derrida and de Man's work. Through this other, "*this* irreplaceable living being that one day enters my space, into this place where it [he] can encounter me, see me, even see me naked," Derrida understands "the absolute alterity of the neighbor" and mortality, the other's and his own: "Nothing can ever rob me of the certainty that what we have here is an existence that refuses to be conceptualized . . . And a mortal existence, for from the moment that it [he] has a name, its [his] name survives it [him]. It signs its potential disappearance. Mine also . . ."[30] For Derrida, naming is "a foreshadowing of mourning because it seems to me that every case of naming involves announcing a death to come in the surviving of a ghost, the longevity of a name that survives whoever carries that name."[31]

Later on in his address (for the text of *The Animal* takes the form of an oral address and bears its traces), Derrida talks about how Cérisy is a haunted place, and names some of the disappeared (though not de Man):

> This chateau has remained for me, for so long now, a chateau of haunted friendship . . . Indeed friendship that is haunted, shadows of faces, furtive silhouettes of certain presences, movements, footsteps, music, words that come to life in my memory . . . I enjoy more and more the taste of this memory that is at the same time tender, joyful, and melancholic, a memory, then, that likes to give itself over to the return of ghosts, many of whom are happily still living and, in some cases, present here.[32]

*Figure 2.* "Avec Lucrèce à Ris-Orangis, vers 1999." © DR. (Source: Cahier Derrida, L'Herne, 2004, used with permission)

On the one hand, absolute alterity; on the other, the singularity, unsubstitutability of that other, the friend, the cat (see figure 2).[33] Derrida's language in this essay echoes the ethical injunctions of another (absent) friend, Emmanuel Levinas, and nowhere more so than when Derrida pointedly distances himself from that other philosopher, in the question "How can an animal look you in the face?," for here he cites an interview where Levinas, pressed on the question of the face of the other, retreated (albeit ambivalently) from conferring on the non-human animal the full dignity and responsibility of a face.[34]

It is in fact in Derrida's discussion of Levinas where the aporia between the literal and the figurative is staged, and where the ethical relation to the other is both questioned and affirmed. In the second part of *The Animal*, Derrida reads in Levinas a symptom of the negation that Derrida himself pursued in arguing that his cat was real.[35] Seeing no fewer than eleven exclamation marks in the space of the eight pages of Levinas's text, Derrida discerns a reversal in the trope of negation: "Moreover, two of them [exclamation marks] follow the utterance 'But no! But no!' which in truth attests to the truth of a 'But yes! But yes!' when it comes to a dog that recognizes the other and thus responds to the other."[36] He concludes that although Levinas insists—as Derrida does above—on the literality of the dog, its specificity, this singular dog with a name, Bobby (Derrida's cat, in this essay, remains nameless), "Levinas's text is at once metaphorical, allegorical, and theological, anthropotheological, hence anthropomorphic . . . at the very moment when Levinas proclaims, claims, *prétend*, by exclaiming, the opposite."[37] Inscribed then in a reading of the brother (the other whose theory of ethical fraternity Derrida carries with him) is a key to the undecidabilities of the singular/general animal, of the "real" cat, in the animal that Derrida is and follows. He reminds us, finally, that "every animal . . . is essentially fantastic, phantasmatic, fabulous, of a fable that speaks to us and speaks to us of ourselves."[38]

And yet Levinas too devoted pages to an animal, a dog with a name, Bobby, a particular, specific dog, "the name of *a* dog" ("nom *d'un* chien"), which is also the polite imprecatory substitute for "name of God," whose disappearance marks the disappearance of all those in the camps who do not, like Levinas, survive to mark their traces.[39] Levinas's essay is rife with the linguistic playfulness and punning of Derrida's, reminding readers of all the ways canine references live and breathe in French; he, too, refers to the Old Testament, to the dogs who collaborated with and conferred dignity upon the chosen people on the night of the death of the first-born in Exodus by failing to bark, thus erasing their traces.[40] Dog is the copilot; dog silence is God's word, God's voice backward.[41] Bobby, that errant dog (a wandering Jew?), enters the lives of the "simianized" "band of monkeys" that the Jews have been forced to become, "beings enclosed within their species; despite all their vocabulary, beings without language," signifiers without a signified.[42] But Bobby, "cherished dog," exotically named by those who loved him, leapt and barked gaily upon their return from hard labor each day and, in so doing, reminded them that they were human (no, rather, that they were *men*) and claimed descent from those silent ancestors on the shores of the Nile.[43] That animal other, he who does not have the capacity to universalize, writes Levinas, confers humanity on the human, enters the human prison from the margins, and calls to the human within.[44] But does he have a face? This is not Derrida's question, neither the cat's friendship with "man" (this is why, perhaps, it is a cat and not a dog), nor the cat's face, but the cat's look, his/her gaze, and the returned look of the human. For Levinas, the significance of the non-human animal resides in the fraternity of men that is also the hallmark of a certain humanity. For Derrida, the non-human animal raises, on the contrary, the question of the other, of femininity, and ultimately of sexual and species difference(s).

Why *this* cat (not on the mat but in the bedroom and the bathroom, following, staring)? In a talk at a conference about the autobiographical animal, Derrida mentions Kofman's *Autobiogriffures*, and there too it is a question of an absent friend, an untimely absence, mourned and invoked as reigning muse of cat-scratch philosophy, for Kofman is the last in the AIDS-quilt-like litany of the names of the dead who haunt Cérisy.[45] Kofman reminds us that there is a veritable lineage of famous cats in history and literature, and that they are the "the animals preferred by many writers, as though there were a particular affinity between the cat and writing, the cat and culture."[46] In *Autobiogriffures* Kofman pursues the *griffes*, claws, and marks of writing with Hoffman's cat Murr; from the first pages, she reminds us of Derrida's once-humanist (but not necessarily) articulation of *archi-écriture* [archi-writing]:

> A certain ethnocentrism "refuses the name of writing to certain techniques of consignment," and admits of the existence of peoples "without writing," "without history," to whom it refuses the name of man. *A fortiori* one would not be able to admit that the animal would have a certain disposition toward writing, a certain disposition toward acquiring a certain writing . . . [But] what if the scrawls of a cat were hieroglyphs? Was not the cat sacred in Egypt, the country where Thoth invented writing?[47]

The ethnocentrism Derrida identified in *Of Grammatology* is also anthropocentrism, as he argues in *The Animal:*

> Let me note very quickly in passing, concerning intellectual autobiography, that whereas the deconstruction of "logocentrism" had, for necessary reasons, to be developed over the years as deconstruction of "phallogocentrism," then of "carnophallogocentrism," its very first substitution of the concept of the trace or mark for those of speech, sign, or signifier was destined in advance, and quite deliberately, to cross the frontiers of anthropocentrism, the limits of a language confined to human words and discourse. Mark, gramma, trace, and différ*a*nce refer differentially to all living things, all the relations between living and nonliving.[48]

Cats write, as the *griffe*—simulacral signature and inscription, trace, mark—suggests. In 1976, Kofman argues that Hoffman (and Tieck before him, in his *Puss N' Boots*) challenges the human/non-human animal distinction precisely around this question of writing. Writing, in the form of the cat's *griffe*, lacerates, shreds the book, confounding inside and outside, authorship, and private property: it "puts human writing in its place as being only one type of writing."[49] This struggle between "man" and animal around the trace, the effacement of the trace, inscription, is also at stake in Lacan's insistence that the non-human animal does not cover up its tracks. In a magisterial reading of Lacan around, precisely, the link between tracing and tracking and their effacement, the undecidability of the antinomic senses of the word *dépister*: to track, to follow a scent or tracks, and to confuse by covering one's tracks, Derrida reminds the reader that "the structure of the trace presupposes that to trace amounts to erasing a trace as much as to imprinting it, all sorts of sometimes ritual animal practices—for example, in burial and mourning—associate the experience of the trace with that of the erasure of the trace."[50] The point is not whether the animal can or cannot erase or efface its traces, but that the human cannot do so either:

> The trace cannot be erased . . . A trace is such that it is always being erased and always able to be erased . . . But the fact that it can be erased [*qu'elle s'efface*], that it can always be erased or erase itself, and that from the first instant of its inscription . . . does not mean that someone, God, human, or animal, can be its master subject and possess the power to erase it . . . In this regard, the human no more has the *power* to cover its tracks than does the so-called animal.[51]

What rejoins Kofman's description of the cat's "innocent" lacerations of the human inscription/trace, and raises the question of usurpation, is Derrida's remark that the distinction between the human and the so-called animal being upheld in Lacan's argument testifies to an "anthropocentric reinstitution of the superiority of the human order over the animal order, of the law over the living," and that "such a subtle form of phallogocentrism seems in its way to testify to the panic Freud spoke of: the wounded reaction not to humanity's *first* trauma . . . but rather to its *second* trauma, the Darwinian."[52] Tracing, tracking, following; inscription, trace: these are the aporetic (and yet not!) paths of Derridean and feline animality, the in- or a-human in the human and non-human animal; the following that is another way of saying "I am," when what "I am" is following the other.[53] Before the question of being as such, there is the question of following (for persecution or seduction).

Derrida does not, like Kofman, focus on the cat's claws, the *griffure/griffe* that is the mark of marking, the animal weapon that also inscribes. His fear, naked before his little female cat, is not a fear of being clawed, lacerated, written on or over by the cat; no, it is a look with which he is concerned, a mouth too that he is—for a moment—afraid of: "The cat observes me *frontally* naked, face to face, and if I am naked faced with the cat's eyes looking at me as it were from head to toe, just *to see*, not hesitating to concentrate its vision—in order to see, with a view to seeing—in the direction of my sex. *To see*, without going to see, without touching yet, and without biting, although that threat remains on its lips or on the tip of the tongue."[54] Who, then, is this cat that looks, that perhaps, at least in Derrida's imagination, thinks about touching or biting, and doesn't, at least not yet?

There is something feminine about the cat. She is a female cat, Derrida says, after calling her a (masculine) cat; her gender will appear and disappear over the course of the essay, for when appearing as example, the cat is masculine (the unmarked gender), but in her occasional singularity, she is feminine. Naked, before Derrida, she is sometimes a female cat. The English translation commits the sin Vicki Hearne denounces along with "scare quotes around animal names," of doing what "even many animal lovers do": using "it" rather than "he" or "she" to refer to an animal, as English rather than French can do to indicate the generic animal.[55] But it is here precisely a question of the non-generic; there is no animal-as-such. Something about sexual difference and something about kinship are troubled when an animal enters the scene. Echoing a passage from *Politics of Friendship*, Derrida says we must ask ourselves "what happens to the fraternity of brothers when an animal enters the scene."[56] He notices that, in thinking through his zoo-auto-bio-bibliography, "animals are welcomed . . . on the threshold of sexual difference. More precisely of sexual differences."[57]

The pluralization of sexual differences queers this animal world on the way to an analysis of human masculine shame: here is a sex that is not one, a sexual difference that is not one either, but many, or more than one.[58] More than one difference appears as well in the room, for the shame that shows up, stands up, heats up we might say with Derrida, when Derrida is in the room with the cat, is a third term:

> The first [difference] is when another is in the room, when there is a third party in the bedroom or the bathroom, unless the cat itself, whatever its sex, be that third party. Allow me to make things still more clear: all that becomes all the more acute if the third party is a woman. And the "I" who is speaking to you here dares therefore to posit himself, he signs his self-presentation by presenting himself as a man, a living creature of the masculine sex, even if he does so . . . retaining an acute sense of the unstable complexity . . . even suspecting that an autobiography of any consequence cannot not touch on this assurance of saying "I am a man," "I am a woman," I am a man who is a woman.
>
> Now this self, this male me, believes he has noted that the presence of a woman in the room heats things up in the relation to the cat, vis-à-vis the gaze of the naked cat that sees me naked, and sees me see it seeing me naked, like a shining fire . . . besides the presence of a woman, there is a mirror [*psyché*] in the room. We no longer know how many we are then, all males and females of us.[59]

The animal in the room, the animal(s) in the "psyche" (or "cheval") mirror, generate sexual differences from sexual difference, even as they institute sexual difference through what Derrida calls "hetero-narcissistic" erotic mirroring, a mirroring of the self as other, a mirror stage that defines the moment of desire *and* identification as a moment of pursuit of the other, seduction, of "seductive pursuit."[60] In his wandering through non-specular animal desire, or rather the nonvisual specularity of animalian mirroring (through sound and scent as well as look) that constitutes recognition of the other, an other of the "same" species, Derrida forges a mirror stage, which is also, as his play on the mirror as psyche is meant to remind us, a *psychic* formation for animals, and finds out what, in shame, seems peculiar to the male of the human species: the erection, the rhythmic tumescence and detumescence of a standing-up over which the human male has little control, and which he thus cannot dissimulate. This "general phenomenon of erection" is also, he argues, part of the process of hominization, of coming to stand upright on two legs as a difference from other animals.[61] There is something in this shame, then, that makes it difficult to meet an animal face-to-face:

> It is in this place of face-to-face that the animal looks at me; that is where I have difficulty accepting that what one calls an animal looks at me, when it looks at me, naked. That this difficulty [*mal*] does not exclude the announcement of a certain enjoyment [*jouissance*] is another question still, but one will understand that it is also the same thing, that thing that combines within itself desire, *jouissance*, and anguish.[62]

Later on in the essay, when he reminds us that the comparison of humans to animals has been a way of degrading Jews and women through the absolute idealization of animals, on the one hand, and their absolute demonization, on the other, Derrida returns to the question of masculinity's shame and the *mal* that is both difficulty and evil.[63] There is a sacrificial scene here, the sacrificial structure and logic subtending phallic heteronormativity that Derrida has described elsewhere, especially in "'Eating Well,'" as "phallogocentrism" and "carno-phallogocentrism":[64]

> (Evil intended, harm done to the animal, insulting the animal would therefore be a fact of the male, of the human as *homo*, but also as *vir*. The animal's problem [*mal*] is the male. Evil comes to the animal through the male.) It would be relatively simple to show that this violence done to the animal is, if not in essence, then at least predominantly male, and, like the very dominance of that predominance, warlike, strategic, stalking, *viriloid* . . . it is the male that goes after the animal.[65]

The circuitous route that this deconstruction of humanism takes, then, opens a path for queerness by following the ways "man," in erecting himself into (pre)dominance over the animal, recognizes and disavows that his "I am" is first of all an "I am following," "I am after" the animal, and that the animal, the *animot*—passion, suffering, passivity, not-being-able-ness[66]—is what is "disavowed, foreclosed, sacrificed, and humiliated" in himself on the way to being human.[67]

This is a new kind of male human animal who stands before "his" cat, discomfited by the face-to-face encounter, aware of the abyss in the gaze between them, but also aware of the subjectivity of the other face. In a moment that reaches for a revision of oedipal subject formation (in its brief critique of the Lacanian mirror stage) and, simultaneously, enacts the implications of a queer ethics of intersubjectivity—queer in its cross-species, hetero-narcissistic, erotic mirroring—Derrida asks,

> And can I finally show myself naked in the sight of what they call by the name of "animal"? Should I show myself naked when, concerning me, looking at me, is the living creature they call by the common, general, and singular name *the animal*? Henceforth I shall reflect (on) the same question by introducing a mirror. I import a full-length mirror [*une psyché*] into the scene. Wherever some autobiographical play is being enacted there has to be a *psyché*, a mirror that reflects me naked from head to toe. The same question then becomes whether I should show myself but in the process see myself naked (that is, reflect my image in a mirror) when, concerning me, looking at me, is this living creature, this cat that can find itself caught in the same mirror? Is there animal narcissism? But cannot this cat also be, deep within her [*sic*] eyes, my primary mirror?[68]

The fullest queering, we could say, happens in a moment of mistranslation, a monolingualism of the other perhaps as well, a monolingual humanism (a *hum*onolingualism?) that marks the cat who looks at Derrida. Marks *her*, to be sure, because Derrida elsewhere designates her as a female cat, "his" female cat, whereas the male cats have all pranced, strolled, or crept through the essay in a neutered English form, the neuter; but also because, in the Lacanian or object relations mirroring this scene recalls, the primary mirror is the (m)other. To make of Derrida a cat's kitten (Alice's too, perhaps) is indeed to queer kinship. But Derrida's desire for "his" cat, his identification with her, does not necessarily belong to such a filial register; it is something other, a queer, ontological, abyssal relation, a relation to the feminine human and also to a non-human other. And to find a mirror in the other male (other) is also to queer the intersubjective erotics of a hetero-narcissism-in-the-making. In all these cases, with all these sexes and sexual differences at play in a hall of mirrors, queer animalian theory "is" "after" Derrida, just as "he" is "after" her.

## POSTSCRIPT

Another ethical theorist, toward the end of his followings-after, meditated, perhaps melancholically, on the ends of "man" and on relations among the living who share the passionate condition of mortality. To end where I began, to add one more *griffure* to Sarah Kofman's feline genealogy, I cannot help but wonder whether Derrida's cats are the interlocutors—no, the mirrors—of a certain other gaze, not this time the gaze of the human-in-the-making, naked like an infant, but that of the philosopher contemplating death. Claude Lévi-Strauss, in a beautiful passage that closes *Tristes Tropiques* (that "sad" place he went to as a Jew fleeing Europe, following the tracks or traces of an earlier denunciator of carno-phallogocentrism, Jean de Léry, who was also fleeing Europe's religio-racist persecutions), writes, of "our" time between the beginning of the world and its end,

> The world began without man and will end without him . . . man is not alone in the universe . . . as long as we continue to exist and there is a world, that tenuous arch linking us to the inaccessible will still remain, to show us the opposite course to that leading to enslavement; man may be unable to follow it, but its contemplation affords him the only privilege of which he can make himself worthy; . . . a privilege coveted by every society, . . . the possibility, vital for life, of *unhitching*, which consists . . . in grasping, during the brief intervals in which our species can bring itself to interrupt its hive-like activity, the essence of what it was and continues to be, below the threshold of thought and over and above society: in the contemplation of a mineral more beautiful than all our creations; in the scent that can be smelt at the heart of a lily and is more imbued with learning than all our books; or in the brief glance, heavy with patience, serenity and mutual forgiveness, that, through some involuntary understanding, one can sometimes exchange with a cat.[69]

## NOTES

Many people helped me find threads of my research and of my thought for this essay; thanks go to Michael O'Rourke, Cora Diamond, Maria Frangos, John Freccero, Noreen Giffney, Wlad Godzich, Jody Greene, Rachel Jacoff, Richard Macskey, Laurel Peacock, Barbara Spackman, Carra Stratton, Helen Tartar, and Andrzej Warminski. I dedicate this essay to my feline companions: Mort, who shadowed my life for a quarter of a century; Trotsky, who accompanied me for only a brief time; Lenin, my beloved companion of fifteen years; and the recently departed Boychick.

1. Jacques Derrida, *The Animal That Therefore I Am*, ed. Marie-Louise Mallet, trans. David Wills (New York: Fordham University Press, 2008). All references in English are taken from this edition. See also *L'animal que donc je suis*, ed. Marie-Louise Mallet (Paris: Galilée, 2006). Page references to the French edition follow the English. For other excerpts of the work, see "The Animal That Therefore I Am (More to Follow)," trans. David Wills, *Critical Inquiry* 28, no. 2 (Winter 2002): 369–418; "And Say the Animal Responded?" in *Zoontologies: The Question of the Animal*, ed. Cary Wolfe (Minneapolis: University of Minnesota Press, 2003), 113–128.
2. Carla Freccero, *Queer/Early/Modern* (Durham, NC: Duke University Press, 2005).
3. Derrida, *The Animal*, 29–31 [*L'animal*, 51–53].
4. For a discussion of the way animal theory might intersect with queer disability studies in articulating the disaggregation of the humanist subject, and extensive bibliography in the field for further exploration, see Sunaura Taylor, "Beasts of Burden: Disability Studies and Animal Rights," *Qui Parle* 19, no. 2 (2011): 191–222.
5. For the term carno-phallogocentrism, see Derrida, "'Eating Well,' or the Calculation of the Subject: An Interview with Jacques Derrida," in *Who Comes after the Subject*, ed. Eduardo Cadava, Peter Connor, and Jean-Luc Nancy (New York: Routledge, 1991), 96–119, at 113 ["'Il faut bien

manger,' ou le calcul du sujet," in *Points de suspension: Entretiens*, ed. Elisabeth Weber (Paris: Galilée, 1992), 269–301, at 294]. See also Alice Kuzniar for a meditation on how intimacy with non-human animals queers relations of desire; "On Intimacy with Dogs: Sensuality, Pleasure, Loyalty, and Love outside the Norms of Heterosexual Relationships," *American Sexuality Magazine* (San Francisco: National Sexuality Resource Center, 2004), http://nsrc.sfsu.edu/HTMLArticle.cfm?Article=754).

6. Donna Haraway, *When Species Meet* (Minneapolis: University of Minnesota Press, 2008), 3. I thank Jody Greene for drawing my attention to this thread in Haraway's thought. Ann Cvetkovich, in *An Archive of Feelings: Trauma, Sexuality, and Lesbian Public Cultures* (Durham, NC: Duke University Press, 2003), also discusses the importance of attending to the afterlife of traumas that could be seen to be uneventful when compared to what are understood to be avowedly historical and public catastrophic events. These are the questions I explore at the end of *Queer/Early/Modern*, not so much to "solve" a problem of temporal accountability as to suggest alternative ways to respond to—and survive—the not strictly eventful afterlife of trauma in a just, queer fashion.
7. Derrida, *The Animal*, 6 [*L'animal*, 20–21].
8. Derrida, *The Animal*, 4 [*L'animal*, 19].
9. Barbara Smuts, "Reflections," in J. M. Coetzee, *The Lives of Animals* (Princeton, NJ: Princeton University Press, 1999), 107–120, at 108.
10. Derrida, *The Animal*, 13–14 [*L'animal*, 31–32]. Derrida, *The Animal*, 40–41 [*L'animal*, 64].
11. Derrida, *The Animal*, 163, n. 8 [*L'animal*, 21, n. 2].
12. Michel de Montaigne, *The Complete Essays of Montaigne*, trans. Donald Frame (Stanford, CA: Stanford University Press, 1957), 328–359, at 331; see "Apologie de Raymond Sebond," *Essais, livre II* (Paris: Garnier Flammarion, 1969), 114–153, at 119. "Se jouer à" has a stronger meaning than being a pastime; it connotes meddling or fooling with.
13. Haraway, *When Species Meet*, 20.
14. See Cary Wolfe on Derrida's discussion of "our subjection to a radically ahuman technicity or mechanicity of language" in "Flesh and Finitude: Thinking Animals in (Post)Humanist Philosophy," in *The Political Animal*, ed. Chris Danta and Dimitris Vardouklakis, a special issue of *Substance* 37, no. 3 (2008): 8–36, at 28. For a discussion of Derridean iterability and dissemination, both of which also point to the inhuman or ahuman of language, signifying, trace, see especially Jonathan Culler, "Philosophy and Literature: The Fortunes of the Performative," *Poetics Today* 21, no. 3 (2000): 503–519.
15. Lewis Carroll, *Alice's Adventures in Wonderland; and, Through the Looking-Glass: and What Alice Found There*, ed. Hugh Haughton (London: Penguin, 1998); Sarah Kofman, *Autobiogriffures du chat Murr d'Hoffmann* (Paris: Galilée, 1984). Kofman's text remains untranslated; all translations are my own.
16. Derrida, *The Animal*, 7 [*L'animal*, 23].
17. Derrida, *The Animal*, 9 [*L'animal*, 26].
18. Derrida, *The Animal*, 3–4 [*L'animal*, 18]. Derrida, *The Animal*, 11 [*L'animal*, 28].
19. Alfred Tarski, "The Concept of Truth in Formalized Languages," *Logic, Semantics, Metamathematics: Papers from 1923 to 1938*, trans. J. H. Woodger, ed. J. Corcoran, 2nd ed. (Indianapolis: Hackett, 1983), 152–278; John R. Searle, "Literal Meaning," *Erkenntnis* 13 (1978): 207–224; J. L. Austin, *Philosophical Papers* (Oxford: Oxford University Press, 1979).
20. See Jacques Derrida, *Limited Inc*, ed. Gerald Graff (Evanston, IL: Northwestern University Press, 1988).
21. For some of Derrida's writing about/on animals, in addition to the texts studied here, see

"Violence against Animals," in Derrida and Elizabeth Roudinesco, *For What Tomorrow: A Dialogue*, trans. Jeff Fort (Stanford, CA: Stanford University Press, 2004), 62–76 ["Violences contre les animaux," in Derrida and Elizabeth Roudinesco, *De quoi demain: Dialogue* (Paris: Galilée, 2001), 105–127]; and "'Eating Well.'"

22. Derrida, *The Animal*, 8 [*L'animal*, 25].
23. Derrida, *The Animal*, 7–9 [*L'animal*, 23–26].
24. Derrida, *The Animal*, 11 [*L'animal*, 28].
25. Paul de Man, *The Resistance to Theory* (Minneapolis: University of Minnesota Press, 1986), 5.
26. Wlad Godzich, "Foreword: The Tiger on the Paper Mat," in de Man, *Resistance to Theory*, ix–xviii. My reading owes a serious debt of gratitude to Godzich's essay.
27. Godzich, "Foreword," xiii. See de Man, *Allegories of Reading: Figural Language in Rousseau, Nietzsche, Rilke, and Proust* (New Haven: Yale University Press, 1979), 150; and Jacques Derrida, *Of Grammatology*, trans. Gayatri Chakravorty Spivak (Baltimore: Johns Hopkins University Press, 1976), 277 [*De la grammatologie* (Paris: Minuit, 1967), 393]. See also Marc Redfield, "De Man, Schiller, and the Politics of Reception," *Diacritics* 20, no. 3 (Autumn 1990): 50–70, at 55–56, for a beautiful discussion of this important passage in de Man's *Allegories*; see also "An Interview with Paul de Man," in *The Resistance to Theory*, 117–118, where de Man talks about his and Derrida's disagreement around Rousseau at the Baltimore conference.
28. Paul de Man, "The Return to Philology," *Times Literary Supplement*, December 10, 1982, pp. 21–26, at 26. In this first publication of the essay, the word "theorist" in the last sentence read "terrorist" instead.
29. De Man, *The Resistance*, 16–17. "Fearful symmetry" comes from William Blake's poem "The Tyger," in *Songs of Experience* (New York: Dover, 1984), 34, line 4. Perhaps de Man was also thinking of what Nikita Khrushchev was supposed to have said in response to Mao's dismissal of U.S. imperialism: "The paper tiger has nuclear teeth"; see "What Are They Fighting About?" *Time* magazine, Friday, July 12, 1963.
30. Derrida, *The Animal*, 9 [*L'animal*, 26]. Derrida, *The Animal*, 11 [*L'animal*, 28]. Derrida, *The Animal*, 9 [*L'animal*, 26].
31. Derrida, *The Animal*, 20 [*L'animal*, 39]. For some of the works on mourning echoed here, see Jacques Derrida, *Memoires: For Paul de Man*, trans. Cecile Lindsay, Jonathan Culler, and Eduardo Cadava (New York: Columbia University Press, 1986); Jacques Derrida, *The Work of Mourning*, ed. Pascale-Anne Brault and Michael Naas (Chicago: University of Chicago Press, 2001); Jacques Derrida, *Politics of Friendship*, trans. George Collins (New York: Verso, 1997); and Jacques Derrida, *Specters of Marx: The State of the Debt, the Work of Mourning, and the New International*, trans. Peggy Kamuf (New York: Routledge, 2006).
32. Derrida, *The Animal*, 23 [*L'animal*, 43].
33. A propos of the unsubstitutable cat, in an interesting moment of ethical agonizing, Derrida asks, "How would you ever justify the fact that you sacrifice all the cats in the world to the cat you feed at home every morning for years, whereas other cats die of hunger at every instant? Not to mention other people?" *The Gift of Death*, trans. David Willis (Chicago: University of Chicago Press, 1995), 71 [*Donner la mort* (Paris: Galilée, 1999), 101]. Thanks to Helen Tartar for drawing my attention to this passage.
34. Derrida, *The Animal*, 7 [*L'animal*, 24]. See "Interview," in *Animal Philosophy: Essential Readings in Continental Thought*, ed. Peter Atterton and Matthew Calarco (London and New York: Continuum, 2004, repr. 2005), 49–50.
35. This chapter is called, in English, "'But as for me, who am I (following)?,'" 52–118. Derrida is

reading Levinas, "Polémiques: Nom d'un chien ou le droit naturel," in *Difficile liberté: Essais sur le judaïsme* (Paris: A. Michel, 1976), 201–202; the English translation used here is "The Name of a Dog, or Natural Rights," in *Animal Philosophy: Essential Readings in Continental Philosophy*, ed. Peter Atterton and Matthew Calarco (New York: Continuum, 2004), 47–50.

36. Derrida, *The Animal*, 115 [*L'animal*, 159]. Derrida's own line about the cat is "No, no, my cat, . . . does not appear here as representative, or ambassador, carrying the immense responsibility with which our culture has always charged the feline race"; *The Animal*, 9 [*L'animal*, 26].
37. Derrida, *The Animal*, 116 [*L'animal*, 160].
38. Derrida, *The Animal*, 66 [*L'animal*, 95].
39. For discussions of this essay, see David Clark, "On Being 'The Last Kantian in Nazi Germany': Dwelling with Animals after Levinas," in *Animal Acts: Configuring the Human in Western History*, ed. Jennifer Ham and Matthew Senior (New York: Routledge, 1997), 165–198; Peter Atterton, "Ethical Cynicism," in *Animal Philosophy*, 51–61; Karalyn Kendall, "The Face of a Dog: Levinasian Ethics and Human/Dog Coevolution," in *Queering the Non-Human*, ed. Noreen Giffney and M. J. Hird (Hampshire, UK: Ashgate, 2008), 185–204.
40. Derrida, *The Animal*, 48 [*L'animal*, 200–201]. On this question of erasing the trace, see Derrida, *The Animal*, 135 [*L'animal*, 185].
41. Derrida, *The Animal*, 48 [*L'animal*, 201].
42. For "monkey talk," see *The Animal*, 49 ["parler simiesque," *L'animal*, 202]; for "gang of apes," see *The Animal*, 48 ["bande de singes," *L'animal*, 201]; for "beings without language," see *The Animal*, 48 [*L'animal*, 201].
43. Derrida, *The Animal*, 50 [*L'animal*, 202].
44. Derrida, *The Animal*, 50 [*L'animal*, 202].
45. Derrida, *The Animal*, 23 [*L'animal*, 43].
46. Kofman, *Autobiogriffures*, 14, my translation.
47. Kofman, *Autobiogriffures*, 10–11, my translation; quotations refer to Derrida's *Of Grammatology*.
48. Derrida, *The Animal*, 104 [*L'animal*, 144].
49. Kofman, *Autobiogriffures*, 77, my translation.
50. Derrida, *The Animal*, 137 [*L'animal*, 185].
51. Derrida, *The Animal*, 136 [*L'animal*, 186].
52. Derrida, *The Animal*, 136 [*L'animal*, 187].
53. Derrida, *The Animal*, 54–55 [*L'animal*, 82–83].
54. Derrida, *The Animal*, 4 [*L'animal*, 19].
55. Vicki Hearne, *Adam's Task: Calling Animals by Name* (New York: Knopf/Random House, 1986), 169.
56. Derrida, *The Animal*, 12 [*L'animal*, 29]. See also: "The fratriarchy may *include* cousins and sisters but, as we will see, including may come to mean neutralizing. Including may dictate forgetting, for example, with 'the best of all intentions,' that the sister will never provide a docile example for the concept of fraternity . . . What happens when, in taking up the case of the sister, the woman is made a sister? And a sister a case of the brother?" Derrida, *Politics of Friendship*, viii [*Politiques de l'amitié* (Paris: Galilée, 1994), 13].
57. Derrida, *The Animal*, 36 [*L'animal*, 59].
58. I am referring to Luce Irigaray, *This Sex Which Is Not One*, trans. Catherine Porter and Carolyn Burke (New York: Cornell University Press, 1985) [*Ce sexe qui n'en est pas un* (Paris: Minuit, 1977)].
59. Derrida, *The Animal*, 58 [*L'animal*, 86].

60. Derrida, *The Animal*, 60 [*L'animal*, 88]. See also Michael Warner, "Homo-Narcissism: or Heterosexuality," in *Engendering Men: The Question of Male Feminist Criticism*, ed. Joseph Boone and Michael Cadden (New York and London: Routledge, 1990), 190–206. Derrida's discussion of "hetero-narcissism" addresses, I think, Warner's concern that male desire for the (male) other is most often scripted as narcissism, that gender difference is understood, in other words, to be difference *tout court*. Derrida's notion of "hetero-narcissism" allows for the simultaneous play of desire and identification, sameness and difference, in meeting the gaze of the other.
61. Derrida, *The Animal*, 61 [*L'animal*, 90].
62. Derrida, *The Animal*, 61 [*L'animal*, 90].
63. Derrida, *The Animal*, 64 [*L'animal*, 93]. See also Derrida, *The Animal*, 104 [*L'animal*, 144].
64. See also, for an elaboration of the sacrificial logic entailed in the passage from "penis" to "phallus," Jean-Joseph Goux, "The Phallus: Masculine Identity and the 'Exchange of Women,'" *differences* 4, no. 1 (1992): 40–75.
65. Derrida, *The Animal*, 104 [*L'animal*, 144].
66. Derrida, *The Animal*, 27 [*L'animal*, 49].
67. Derrida, *The Animal*, 113 [*L'animal*, 156]. See Derrida, *The Animal*, 105 [*L'animal*, 145], where Derrida gently chides Élisabeth de Fontenay for her critique of rationalist humanism's discourse on the animal by pointing out that both reason and humanism also narrowly circumscribe the human. For an analysis that arrives at some similar conclusions, but with specifically feminist vegetarian goals in mind, see Carol Adams, *The Sexual Politics of Meat: A Feminist-Vegetarian Critical Theory* (New York: Continuum, 2000). For analyses of the comparisons between denigrated racial and sexual categories and animals, again in very different traditions of critique, see Marjorie Spiegel, *The Dreaded Comparison: Human and Animal Slavery* (New York: Mirror Books, 1996); and Joan Dunayer, "Sexist Words, Speciesist Roots," in *Animals and Women: Feminist Theoretical Explorations*, ed. Carol J. Adams and Josephine Donovan (Durham, NC: Duke University Press, 1995), 11–26. Giorgio Agamben, in *The Open*, also argues that it is the humiliated "animal within" against which the human defines himself (though Agamben is not attentive to the specificity of the gender of the human so erected); *The Open: Man and Animal* (Stanford, CA: Stanford University Press, 2004).
68. Derrida, *The Animal*, 50–51 [*L'animal*, 76–77]. The French reads: "Mais *ce* chat ne peut-*il* aussi être, au fond de ses yeux, mon premier miroir?" (Italics mine).
69. Claude Lévi-Strauss, *Tristes Tropiques*, trans. John Russell (New York: Atheneum, 1965), 413–415 [*Tristes Tropiques* (Paris: Plon, 1955, repr. 1984), 495–497].

# Paternalism or Legal Protection of Animals?

## *Bestiality and the French Judicial System*

MARCELA IACUB

*Translated by* VINAY SWAMY

Even an admittedly summary analysis of legal rules can lead us to distinguish two different forms of paternalism. The first seeks to punish certain behaviors considered harmful that individuals inflict upon themselves. In so doing, the state limits the individual's powers of self-regulation, extending an old legal tradition that dates back to the late Middle Ages. Critiques of such measures mostly focus on the fact that, for the most part, they do not recognize that people live or die according to their own values. A critique of paternalism is thus also a reminder of the axiological pluralism of liberal democratic societies. The second form of paternalism penalizes private or public expression of a subject's so-called anti-social, even "evil," inclinations. These behaviors are stigmatized juridically not because they are deemed to be dangerous for individuals themselves, but for others in a more or less distant future. This form of legal paternalism is not subject to the same type of critique as the first. It is not the imposition of constraints on values that is denounced, but a transgression of the spheres or domains of government that are seen as germane to liberal democratic regimes. The state, they say, should not care about saving the souls of its subjects, as would a good pastor; such is not its function. And let us note that it is totalitarian regimes that seek to track deviance in the most private and intimate spheres in the name of religion or antidemocratic ideology.

This second form of legal paternalism, which could be called "pastoral," is more difficult to justify in democratic societies because, more than the former, it immediately evokes the power of theocratic states. Thus, such measures use rationalizations or justifications that seek to integrate them into modern penal policy. It is this form of pastoralism that I would like to explore by analyzing a court case that moved French lawyers at the highest levels: the case of a pony named Junior, who was "sodomized" by his master, Gérard X.

## THE STRANGE CASE OF JUNIOR, THE PONY

Alerted by some unknown informant of the existence of this unusual affair, the Society for the Protection of Animals (SPA) and the Brigitte Bardot Foundation denounced Gérard, an employee of the penal system, for "sexual abuse" committed on a domestic, tame, or captive animal. It certainly seems strange to create legal terms for fierce passions. Of what legally objectionable infamy did the SPA and the Bardot Foundation accuse Gérard X? What acts committed on an animal are in fact defined by the offense of "abuse of a sexual nature"?

This offense contained in Article 521-1 of the Penal Code, which also punishes "serious abuse" and "acts of cruelty," had just become law in March 2004. Until then, French law did not contain provisions for specific punishment of a violent act of this nature carried out on animals living under human authority. Admittedly, this does not mean that judges did not condemn those who tormented animals with their sexual organs. Quite the contrary.[1] But to do so, they used the more generic crimes of "serious abuse" or "acts of cruelty" contained in Article 521-1 of the Penal Code. They could also appeal to the infraction of "mistreatment," which provided for milder punishment than the first two options.[2]

Insofar as the existing indictments were sufficient to criminalize such behavior, why was there a need for the further qualification of "sexual abuse"?[3] What good is new legislation that neither adds anything to, nor transforms existing laws? To this we may reply that the new law, which qualified the term "abuse" with the words "of a sexual nature," meant nothing more than the fact that adding "sexual nature" to the abuse inflicted on an animal rendered it necessarily severe. Indeed, French law distinguishes between, on the one hand, crimes of "serious abuse" and "acts of cruelty," and, on the other, a violation for "mistreatment" of animals that is supposed to cover acts less sadistic or brutal than the former. While the first two offenses are punishable with two years of imprisonment, mistreatment is only punishable by a fine.[4] Certainly, the fact of having specified that among all the atrocious forms that brutality against animals may take, those that are sexual in nature are sufficiently serious to be punished as crimes and not as infractions was already quite remarkable, and could have aroused some suspicions about the meaning that the judges would give the new provision. We know that sometimes judges are not content to stick to the letter of the law, and that they try to fathom the almost unconscious desire of the legislator, especially when it comes to certain criminal policies considered to be a priority.[5]

If we followed the letter of the new law that punishes "abuse of a sexual nature," relations with an animal must be accompanied by brutality or violence of any kind, to be punishable. Otherwise, this text should simply have punished any sexual act with an animal, as was the case during the *Ancien Régime*.[6] We must not forget that this new legislation was not meant to penalize people for their sexual deviance, but for the severe suffering inflicted upon a domestic or tame animal, or one held in captivity.[7] If that was not the conscious intention of the legislators, the law should have been placed in the chapter on ordinary sex offenses, that is to say, those that are practiced by humans on other human beings, or even one that is devoted to the protection of minors. Having made these clarifications, let us return to the two protagonists of our story and to the fate that their unsympathetic judges reserved for them.

Gérard X admitted to the criminal court of Chaumont that he had penetrated his pony, Junior, with his penis, but he believed that the act could not be considered to be any form of abuse, because of the anatomical configuration of the animal species to which the pony belongs. Indeed, if Junior had been a chicken, a small dog, or a rabbit, the same act could be considered an abuse. But since his pony was indeed a pony, what difference could there be between sexual penetration and a caress, a pat, or a walk?[8] We know that some people torment small animals to death by sexually penetrating them. In a text condemning these barbaric practices, Carol Adams gives the example of those who behead poultry in order to intensify the convulsions of their sphincters as they are penetrated.[9] The new law could not equate the acts perpetrated by Gérard X on his pony with those of such torturers.

Yet the court did not hear it that way, and Gérard X was sentenced on September 5, 2005, to a term of one-year sentence that was suspended, and what is worse, to be separated forever from Junior, who was to be given to an association that is supposed to love animals appropriately. Gérard X was also sentenced to a total ban on raising any animal in the future and to pay 2,000 Euros in damages to the plaintiffs, the SPA, and the Brigitte Bardot Foundation. Not satisfied with this unjust decision, which violated, in his view, the letter of the new law, Gérard X appealed. To his chagrin, the Court of Appeals in Dijon upheld the decision of the trial court on January 27, 2006. The rationale for that decision deserves some attention here. The magistrates declared: "the accused had carried out acts of sodomy on Junior, the pony which he owned, and had acknowledged that it was a game." It is noticeable that the judges used the word "sodomy" instead of describing the precise acts of which Gérard X was accused. It is also of note that the pony in question was singled out by its sweet name. The judges did not say that this man had inserted his penis into the anus of a pony but that he had sodomized Junior, thus creating between them a sort of dramatic tension closer to that of pedophilia than zoophilia.

The word "sodomy" also implicated the animal's engagement with Gérard X's enterprise, that is to say, to assume that these actions had a sexual significance for both man and pony. One can certainly use terms like this one with regard to minors of a very young age, who do not understand the meaning of certain sexual acts. However, unlike animals, they are able to comprehend at a later date, and sometimes in a precise manner, what happened to them and, as such, to render such acts of remembrance into traumatic events. The animal, on the other hand, is incapable of comprehending *a posteriori*, and thus, of transforming its memory of an act of this nature into a source of particularly intense psychological suffering.

For the court in Dijon, the nature of sexual abuse was not about the processes employed by the perpetrator, but on a specific intersubjective contact between him and his victim. Furthermore, the appellate court attempted to justify why such an act not only implicates sexuality but also the category of abuse. To do this, it declared the following: "These acts—suffered by the animal, which could not exercise any will whatsoever, nor avoid that which was imposed on it and was thus transformed into a sex object—constituted abuse within the meaning of section 521-1 of the Penal Code . . . the excuse that it was a game was not admissible."

According to the court, as in sexual assault between humans, the existence of abuse is justified by the absence of consent of the victim. Yet, unlike humans, the pony cannot consent to anything, and this inability to consent is not temporary, as is the case with minors, or adults incapacitated by an accident or illness. If Junior is an animal and not a human being, it is precisely because of the impossibility we attribute to animals of exercising clearly rational choices. For animals in the category under Article 521-1 of the Penal Code—that is to say domestic, tame, or captive animals, living under human authority—everything is imposed. They do not choose to be transformed into food, to be subject to experimentation, to entertain, or even, when they are lucky, to become pets. The Penal Code merely prohibits individuals, including owners themselves, to impose on animals certain harsh and painful treatments considered "unnecessary" for human purposes. But even if our legislation were to create affirmative obligations that would help animals to live better, as is the case in some countries, this has nothing to do with the choices animals are meant to exercise regarding their own destiny.

It may be argued that minors do not consent to most things that are done to them, and that it is precisely for this reason that they should not be sexually abused: that sexual relationships imposed on them by adults are very clearly not consensual. However, if we protect children from such abuse and from many other types besides, it is because one day they can, unlike animals, give consent, and not just in the domain of sexuality. Minors are considered to be under adult care, to be educated and socialized until they can become masters of their own destiny (and perhaps of those of dogs or cats as well). If we affirm that children cannot give consent, it is because their ability to consent in the future is given great importance. In contrast, the animal's non-consent has nothing to do with such an expectation for its future life. Saying that an animal is not consenting is redundant if the sentence intends to describe what we mean by "animal." Furthermore, from being redundant, it becomes false if it is used only in the light of a single act, for, by contrast, it suggests that other acts have resulted from explicit consent.

Dissatisfied with the ruling, Gérard X tried to find relief from the Court of Cassation [Supreme Court]. This court's mission is to provide a uniform interpretation of the law. Thus, Gérard could expect that the new judges would interpret the expression "abuse of a sexual nature" in a more reasonable manner. In his appeal, Gérard X put forward that "the sexual penetration of an animal by a human penis, called the act of zoophilia," could not be classified as abuse of a sexual nature in "the absence of violence, abuse or ill treatment." However, the Court of Cassation was not convinced by this interpretation of the law. In its September 4, 2007, decision, it specified that "acts of sexual penetration committed by a person on an animal constitute abuse of a sexual nature" as provided for in the Article 521-1 of the Penal Code. Without adopting the arguments of the Dijon Court, which relied on the non-consent of Junior, the Court of Cassation ruled that the mere act of sexually penetrating an animal constitutes the offense, without the courts having to characterize thereafter the violence, brutality, or mistreatment suffered by the animal.

Commentators on this decision have questioned if the meaning of the concept of "abuse of a sexual nature" was meant only to include penetration, or if it could include other acts, such as being penetrated by an animal, performing oral sex, or sexual fondling. Most scholars have concluded that given the nonviolent nature of the sexual penetration of the pony, there was no reason for the definition of abuse not to extend to acts other than penetration with a human penis.[10] In short, the court reintroduced into French law the crime of zoophilia or bestiality, even if that prohibition is not absolute given that it does not include wild animals but only those that live under human authority. Indeed, after this decision we can legally, as before, engage in all kinds of sex with a wild pig, for chastity should apparently only preside over humans' relations with farm-raised pigs or those that are so sadly subjected to industrial farming. Before taking into account more carefully the problems that the court's decision raises in light of the legal principles and policies of the sexual revolution—in the name of which all sexual offenses between humans have been constructed—we should briefly consider this decision in light of the law that protects animals in France.

## OF THE SEVERITY OF THE DAMAGE SUFFERED BY JUNIOR ACCORDING TO THE FRENCH COURTS

To understand the scope of the decision of the Court of Cassation, which is to say that a nonviolent sexual act with an animal should be punished in the same manner as serious maltreatment and acts of cruelty, let us recall the type of acts that French jurisprudence designates as such. Sodomizing Junior, the pony, was considered *as serious* as the following acts: "organizing a competition for ratter dogs whose snapping jaws supposedly put to death instantaneously rats captured from landfills and held in a cage of appropriate size. In reality, among the animals sacrificed, many die after prolonged agony";[11] "leaving a dog unattended for forty-eight hours, when the animal had its throat torn as a result of two shots fired by the accused";[12] "castrating a horse without anesthetic";[13] "suspending live lambs from hooks by piercing their legs";[14] "kicking and punching a three-month-old puppy, causing multiple fractures";[15] "capturing a cat, sealing it in a bag, and killing it by throwing the bag violently to the ground";[16] "having killed a cat by putting it in a freezer."[17]

It also seems useful to recall the rather violent situations that were *not* held by the judges as brutal enough to constitute the crime of "serious abuse" and "acts of cruelty." Thus, for instance, French judges considered the following behaviors as *less serious* than the act committed by Gérard X on his pony: "having left cows and horses in a pasture without food or drinking water when the weather was unusually severe";[18] "having left a dog in the sun, tied up with a choke-chain, which resulted in his death";[19] "having proceeded, following an ancestral tradition, to slaughter dogs for consumption, although not only could one not establish any evidence of intense suffering during drowning, but that it was also strictly necessary to ascertain, in order for them to be sacrificed for consumption."[20] Furthermore, this same section 521-1, in the name of which Gérard X was penalized, allows both cockfighting and bullfighting,[21] although these brutal acts against those animals are not justified by any worthwhile need for humans, such as the need to eat or to conduct scientific experiments.

In addition to such bloody games, France permits, as we know, the production of *foie gras.*[22] If, of all current and massive abuses to which animals are subjected in the world today, we have chosen just this one, it is largely because the practice is increasingly challenged in democratic countries and because, as in the case of Junior, the pony, it involves penetration into the bodies of the victims, which might contextualize the importance and impact of the so-called sexual act. To produce *foie gras*, every year, 37 million ducks and 2 million geese are force-fed in France. The technique consists of inserting a 20- to 30-centimeter-long pipe down the throat of the animal into its stomach in order to administer a large amount of food they would not have eaten on their own. The operation lasts forty-five to sixty seconds with the traditional feeding method, and two to three seconds to force-feed by a modern pneumatic pump, which can fill more than 350 ducks an hour. This takes place twice a day. It is the equivalent of forcing seven kilos of pasta in a few seconds, twice a day, directly into the stomach of a man weighing 70 kilograms. In due course, the animals suffer from lesions, throat pain, stress, diarrhea, and difficulty in breathing. At the end of the feeding regime, breathing and movements become difficult because the animal's pulmonary sacs are compressed and crushed by the liver. To better carry out these nonsexual "penetrations," 80 percent of ducks are completely immobilized in individual cages with only their heads jutting out, in almost total darkness.[23]

These few examples (and the list could go on) show that it is unlikely that French judges

*Figure 1.* An anti-bullfighting demonstration in Nîmes, France. The definition of animal abuse in the penal code does not extend to the violence inflicted during a bullfight; this protester's sign is calling specifically for a reform of the penal code. (Source: citizenside.com, image "Manifestation anti-corrida," used with permission)

intended to punish Gérard X for the supposed plight of Junior. The suffering inflicted on animals has to be very brutal for Article 521-1 of the Penal Code to be applied, while this law authorizes other kinds of unnecessary suffering, such as bullfighting and cockfighting, and such contestable activities as the production of *foie gras* in France. Therefore, rather than exhibiting concern for the welfare of the animal, the sentencing of nonviolent zoophilia seeks to prohibit humans from expressing abnormal or perverse desires.

It is quite possible that all of this legislation, and not just the crime of sexual abuse, is less concerned with the welfare of animals—to which humankind has unfettered access—than with the individual abuses inflicted upon them that are uncontrolled by the collective. It is as if the cruelty that our societies inflict on animals had to be socially determined, that it should be born of desires turned into necessities by law, rather than of purely private whims. It is therefore quite possible that by punishing those who abuse animals for personal pleasure, we are justifying, above all, the large-scale organized violence perpetrated by society. But in so doing, we have also sought, since the first so-called animal protection laws passed in the nineteenth century, to exercise a certain control over an individual's unhealthy and perverse inclinations.[24] The animal not only exists to be eaten, to be experimented upon, to produce profits, and to ease our loneliness, but also serves as a space for the expression

*Figure 2.* FORCE-FEEDING OR "GAVAGE" OF A GOOSE IN THE PRODUCTION OF *FOIE GRAS*. THE PHOTO SHOWS A MORE TRADITIONAL *GAVAGE*: MODERN METHODS OFTEN INVOLVE IMMOBILIZING THE BIRDS IN TINY CAGES FOR MOST OF THEIR LIVES. THIS KIND OF "PENETRATION" IS ENTIRELY LEGAL UNDER FRENCH LAW, AND TAKES PLACE MILLIONS OF TIMES A DAY IN FRANCE, WHEREAS THE PENETRATION OF A PONY FOR PLEASURE IS CONSIDERED IN A VERY DIFFERENT LIGHT. (SOURCE: WIKIMEDIA COMMONS, HTTP://EN.WIKIPEDIA.ORG, AUTHOR JÉRÔME S., USED WITH PERMISSION)

of internal antisocial tendencies and forms of reprehensible enjoyment that the state, like a good shepherd, seeks to condemn. Everything is beneficial in animals; nothing is thrown away. Everything is transformed into utility, including political utility, that is to say, into a tool to govern human beings.

Of all the moral inclinations the expression of which is objectionable under this legislation for the protection of animals, none is considered more serious than that which draws sexual gratification from an animal. And if we can say this, it is because the interdiction of such sexual pleasures, unlike other reprehensible pleasures, does not require the animal to have suffered any pain.

The hierarchy of evils was undoubtedly disrupted in the law that regulates human relations through sexual offenses voted into law from the mid-1970s in democratic countries. The laws arising from the sexual revolution, which had advocated the liberation of desires and pleasures, have become, somewhat paradoxically, the most repressive ones that Western history has ever known.[25] Without doubt, the most disturbing character of this new law is to have virtually reversed the old hierarchies between blood crimes and sex crimes to the point that, nowadays, the latter are punished and condemned more than the former. In France, according to statistics, rape is punished more severely than the murder of a fifteen-year-old.[26]

In most democratic countries, sexual offenses have become the epitome of crimes in recent decades. There seems to be no change in this trend.[27] On the contrary, it has instead spread and invaded many new fields, as is the case with "virtual" child pornography,[28] as well as in international law offenses.

In this context, it is impossible not to make links between the conviction of Gérard X and the ways in which our society punishes sexual offenses between humans. The nature of such links remains to be understood. For this, we must return to the extremely laconic decision handed down by the Court of Cassation that we left open earlier.

## HOW THE LAW FOR THE PROTECTION OF ANIMALS ALLOWS US TO UNDERSTAND WHAT IT MEANS TO RAPE A PERSON

The reactions of many legal scholars to Gérard X's conviction were harsh and severe, not out of sympathy for those deviant desires that demean man, some say, to the level of the beast,[29] but rather because of the problem of coherence that the conviction poses with regard to the law on animal protection. According to these critiques, the Court of Cassation's decision had transposed into legislation a new sexual offense whose goal was to protect humans by putting on hold laws that actually protect animals; the ruling was thus self-contradictory because not intended to protect animals against the suffering inflicted upon them.[30] In their eyes, the law, and especially the Court of Cassation's decision, can only be explained by the preventive and safety-first logic of sex offenses in which small acts such as indecent exposure, for example, are taken to be the first step in a path that inevitably leads to the rape and murder of a child.[31] In short, the criminalization of nonviolent zoophilia ought not to belong in the legislation to protect animals. Rather, it could be an additional legal tool to protect people against sexual predators.

However, if such had been the raison d'être of the law, as I mentioned earlier, it should have penalized zoophilia in a chapter devoted to attacks on persons, and have submitted offenders to post-penal monitoring that is provided for the perpetrators of crimes and sexual offenses. In addition, the law ought to have punished these acts not only with domestic or tame animals, or those that are held in captivity, but also with wild animals. Although they are related, therefore, nonviolent zoophilia and sexual offenses do not belong in the same legal framework. A second possibility would be to suppose that the Court of Cassation did not introduce an extraneous element into the animal protection law, but had instead tried to effect a kind of translation into animal-related legislation of sexual offenses between humans. This goal seemed to be indicated by the new 2004 law, which singled out, for the first time, the "sexual nature" of the abuse of an animal.

A principle capable of allowing for such a translation had to be found. Such a principle was to express the reasoning underlying the convictions for sexual crimes and offenses between humans. In other words, the court had to analyze and identify the rational basis of existing sexual offenses. Why do we punish sexual violence? What do we punish in these violent acts? This first step was essential to effect a translation. Under this condition, we could develop a rule in the animal protection law to account for everything that is condemned both in sex offenses and in those that govern the relationship between humans and animals.

The court's task was not easy given that modern sexual offenses cannot be conceived without the idea of "violation of consent" and the psychological or moral suffering that accompanies such an act. This idea, or rather this principle, is not a trivial matter. Not only is all modern sexual legislation (abortion, equality between illegitimate and legitimate children, legality of homosexual relationships and couples, etc.) indebted to it, but it is also mostly in its name that the severity of penalties applied to sex offenders has been justified.[32] And the psychological damage that is attributed to these offenses is supposed to result directly from the breach of consent. To translate the spirit of this legislation into the law for the protection of animals was, to begin with, very difficult because animals are beings that are ontologically deprived of consent. Physical pain should have been kept as a criterion because it can be said that among living beings devoid of freedom, sensitivity to pain can take the place of moral injury in humans.[33] It is in these terms that most animal advocates express the common nature that binds animals to humans. However, as we have seen, this is not how the court effected its translation. It used the term "sexual penetration committed on an animal" to define the new offense, a variant of the definition of rape given in Article 222-23 of the Penal Code.[34] To define "abuse of a sexual nature," the only thing we need is the mere existence (and no more) of a sexual act, which, without being agreed to, cannot be subject to consent. It is perhaps for this reason that the legal scholar Michel Véron has noted with irony, in his comments on this decision, that it "thus results in a double assessment of the act of sexual penetration committed by a human being, depending on whether the victim is a human being or an animal. In the first case it is punishable only if committed with violence, coercion, threat or surprise. In the second case, the penalty is imprisonment without the need to establish the use of these means."[35] Suffice it to say that violation of consent plays no part in this definition, let alone the suffering caused to the victims.

What is left, then, of the "nature" of sexual offenses between humans in this definition? What remains is something that seemed to be purely secondary, meant to protect that which is most important, that is, the victims' consent: the disapproval of the expression of deviant desires by the aggressors. By this, they mean unhealthy and perverse urges that the aggressors externalize when they commit such acts. In this logic of the court, what constitutes the essence of sexual offenses would be the penalization of criminals' antisocial inclinations, not the violence or damage they inflict on their victims.

If this interpretation of the court were correct, if it translated the spirit underlying sexual offenses, we ought to consider the fact that to penalize virtual child pornography, for example, is not a method for preventing more serious acts. If we penalize such acts, it is because the basic transgression and the fundamental interdiction—that which the law condemns above all—is the expression of certain moral vices. The harm inflicted on people would therefore only aggravate this transgression instead of constituting it as such. We might as well say that rapists should be punished because their desires are filthy and repugnant to us, and not because they inflict harm on their victims.

If this interpretation were correct, zoophilia would bring us back to a curious synthesis of crimes engendered by the sexual revolution. It would show us that, far from having created offenses based on individual freedom, the sexual revolution seeks to police our vices carefully. Is a legal rationale obsessed with moral purity, and less concerned with penalizing violence against persons, not the most vulnerable to excessive immoderation, and to disparity between acts and penalties? Is it not bound to investigate crimes and offenses in the greatest depths of the human soul?

Such an image of our sexual modernity inevitably evokes the *Ancien Régime*'s ways of governing sexuality, which, under the pretext of impurity, sought to heal the souls of sinners by fire rather than punish sexual behavior harmful to other individuals. Yet it would be very unfair to think that there is no "progress" in history, as the apologists of the sexual revolution like to affirm. For if Gérard X had sodomized Junior in the seventeenth century, he would have been burned at the stake, along with his poor beast.

## NOTES

1. In particular, see the case of the Cour d'appel (CA) of Rouen, 20 November 2000: JurisData no. 2000-039065, in which the accused was tried for the "rape" of a sheep.
2. Translator's note: the three legal categories, in French, are respectively "sévices graves," "actes de cruauté," and "mauvais traitements."
3. Translator's note: "sévices sexuelles" in French.
4. For a history of this legislation in France, see Jean-Pierre Marguénaud, *L'animal en droit privé* (Paris: PUF, 1992); and Jean-Yves Maréchal, "Sévices graves ou actes de cruauté envers les animaux," *JurisClasseur Pénal*, arts 521-1 et 521-2: fasc. 10 (Paris: LexisNexis SA, 2009).
5. This is so, notably, on matters of criminalizing sexual offenses not only in France but also in other countries such as the United States. In this regard, it is interesting to note Friedrich Hayek's observations on the role of judges in interpreting the law with regard to legal revolution: "Every lawyer, when called upon to interpret or apply a rule which does not accord well with the rest of the system, strives to arrive at a reading such that it is consistent with other rules . . . The situation was completely different, however, when a philosophy of law that was not in agreement with the majority of existing laws, recently prevailed. The same lawyers, with the same habits and techniques, became generally as efficient in transforming the law, in its every detail, as they were before in keeping it unchanged. The same forces that, in the first situation, resulted in the absence of movement, became, in the second, a factor for accelerated change until the whole body of law was transformed to a point that no one anticipated or desired." *Droit, Législation et Liberté*, foreword by Philippe Nemo (Paris: PUF, 2007), 176–177.
6. From the fifteenth century on, the crime of bestiality was subject to sensational trials, which ended with the condemnation of the accused man or woman and animal to be burned at the stake. See Ludovico Hernandez (Fernand Fleuret), Les procès de bestialité au XVIe et XVIIe siècles: Documents judiciaires inédits publiés avec un avant-propos (Paris: arcanes, 1955). The Revolutionary Code of 1991 put an end to this repression thought to be barbaric, archaic, and religious in nature.
7. Many countries, including Australia, Canada, the United States, and the United Kingdom, prohibit nonviolent bestiality. It is permitted in Denmark, the Netherlands, and Sweden. Germany, Belgium, and Russia only prohibit it for use in the pornography industry.
8. For a defense of nonviolent bestiality from the perspective of utilitarian philosophy, see Peter Singer, "Heavy Petting," *Nerve*, March/April 2001.
9. Carol Adams, "Bestiality: The Unmentioned Abuse," *Animals' Agenda* 15 (1995): 29.
10. In particular, see Jean Yves Maréchal, "Sévices graves," as well as Jérôme Chacornac, "La définition sur mesure d'une infraction à la finalité incertaine," *Dalloz, Etudes et commentaires* 8 (2008): 524–527.
11. Douai, 15 July 1983: *Gaz Pal.* (*Gazette du palais*) 1983. 2. 540.
12. Paris, 16 October 1998: *Droit pénal* 1999. 51, obs Véron.

13. Pau, 24 April 2001: *JCP* 2001.IV. 3102.
14. CA Bordeaux, 11 March 1986: *JurisData* no. 1986-041054.
15. CA Angers, 22 November 2007: *JurisData* no. 2007-362536.
16. Cass. Crim. 4 February 1998: *JCP* (*Juris-classeur périodique*) *G* 1998. IV. 2027.
17. CA Pau, 28 April 2005: *JurisData* no. 2005-274606.
18. Cass. Crim. 23 January 1989: *Bull crim* (*Bulletin des arrêts de la Cour de cassation, Chambre criminelle*) 1989, no. 23.
19. CA Nîmes, 26 September 1997: *JurisData* no. 1997-030522.
20. Papeete, 19 February 1998: *Droit Pénal* 1999.51, obs Véron.
21. For a critique of the brutality of bullfighting and the arguments in its defense, see Jean-Baptiste Jeangène Vilmer, *Ethique animale*, preface by Peter Singer (Paris: PUF, 2008), 210–217. For a defense of bullfighting, see Francis Wolff, *Philosophie de la corrida* (Paris: Fayard, 2007).
22. The need to produce *foie gras* is increasingly challenged in democratic countries. Force-feeding is banned in Germany, Luxembourg, Norway, Sweden, Finland, Denmark, Czech Republic, Netherlands, Poland, Switzerland, United Kingdom, Italy, Austria, Turkey, and Israel. In California, the prohibition came into effect starting in 2012.
23. See Jean-Baptiste Jeangène Vilmer, *Ethique animale*, 179.
24. For an excellent analysis of the origins of the legislation for the prevention of cruelty to animals, as relating to the inclinations of men in violent crimes, see Maurice Agulhon, "Le sang des bêtes: Le problème de la protection des animaux en France au XIXè siècle," in *Si les lions pouvaient parler: Essais sur la condition animale*, ed. Boris Cyrulnik (Paris: Gallimard, 1998), 1185–1219.
25. See Marcela Iacub, *Par le trou de la serrure: Une histoire de la pudeur publique* (Paris: Fayard, 2008); and *Le crime était presque sexuel et autres essais de casuistique juridique* (Paris: Champs Flammarion, 2003).
26. See Marcela Iacub and Patrice Maniglier, *Antimanuel d'éducation sexuelle* (Paris: Bréal, 2005).
27. For the United States, see, among others, Debbie Nathan and Michael Snedeker, *Satan's Silence: Ritual Abuse and the Making of a Modern American Witch Hunt* (New York: Author Choice Press, 2001).
28. See Marcela Iacub, *De la pornographie en Amérique: La liberté d'expression à l'âge des démocraties délibératives* (Paris: Fayard, 2010).
29. Alain Sériaux, JCP 2007, special issue, *cahier* 2: 75.
30. For this perspective, see Jérôme Chacornac, "La définition," and Jean Yves Maréchal, "Sévices graves."
31. For the problem of exhibitionism and its relation to committing more serious crimes, see Marcela Iacub, *Par le trou de la serrure*.
32. See in particular Marcela Iacub and Patrice Maniglier, *Antimanuel d'éducation sexuelle*. See also Francis Caballero, *Droit du sexe* (Paris: LGDJ, 2010).
33. On this issue, see, among others, the French translation of Peter Singer's *Animal Liberation*, *La libération animale*, trans. Louise Rousselle (Paris: Grasset, 1993).
34. "Any act of sexual penetration of any kind whatsoever, committed against another person by violence, coercion, threat or surprise, is rape."
35. Michel Véron, *Droit Pénal, commentaires*, no. 133, November 2007.

# PART 4

# Animals and Environment

# On Being Living Beings

## *Renewing Perceptions of Our World, Our Society, and Ourselves*

ISABELLE DELANNOY

*Translated by* MARIÈVE ISABEL

It may seem curious to answer a call for papers on the animal question in French thought by proposing an article about an awareness that goes beyond the animal, an awareness of being living beings that happens, in my view, through aesthetic experience. This is due on the one hand to my training as an agricultural engineer, with a background in the sciences where students learn to apply techniques to transform the living world,[1] and on the other hand to my experience as an environmental activist, where part of my professional life has consisted in popularizing environmental issues by combining images and text. Drawing on this experience, I will argue in this paper that there is an emerging awareness, at collective and global levels, of being living beings, something others have called a "biospheric"[2] or "ecocentric"[3] awareness. I will explain why this collective awareness of being living beings is an essential element for creating new pathways that will lead us out of the dead end of our present model of development. I will argue that aesthetics have a major role to play in this transformation that invites us to move beyond certain zoocentric ways of viewing the world. In the first section of the article, I will show how this awareness of being living beings has grown since the 1960s. While doing so, I will identify some of the main aesthetic components of this awareness. In the second part, I will explain what links this growing ecological awareness to the animal question, what differentiates it from the animal question, and how this ecological awareness invites us to consider the animal question from a wider angle. Lastly, I will show that many individuals in different fields (scholars, activists, entrepreneurs, etc.) are currently inventing new ways of understanding humanity as both an integral part of the biosphere and at the same time a creative being within it. I will argue for an aesthetics of the ecological question, the growing awareness of which is acting as a powerful catalyst for the emergence of this new collective paradigm.

## PERILS ON THE RISE: AN UNPRECEDENTED ECOLOGICAL CRISIS

Since the 1950s, human activity has modified the ecosystems of the earth faster than ever before in human history. The quality of soil has decreased on over a quarter of agricultural land, the composition of the atmosphere is changing, and global temperatures are on the rise. It is now possible to assert that the average temperature of the earth's surface will increase by 2°C in the next decades, which could trigger an acceleration of global warming through positive feedback loops. This could mean an increase of 4°C by 2060 and 6°C by the end of the century,[4] which would lead us into a world as yet unknown in human history. The oceans' equilibriums are being modified. The extinction rate of species is 100 to 1,000 times greater than natural extinction rates, as recorded in the earth's geological layers, and this phenomenon could increase by 1,000 to 10,000 in the near future. It is not an exaggeration to say that these ecological transformations are challenging humanity with issues never before seen. The human being is now a major geological force on the globe. Our technological power and our ability to transform resources are a major cause of the changes we are observing. The question we must now ask is whether we can change our models and, more generally, the ways of thinking that shape these models.[5]

## THE RISE OF THE GREAT ECOLOGICAL NARRATIVE AND ECOCENTRIC AWARENESS

Paralleling the rise of these perils, an ecological consciousness is growing at the collective and global levels. Sociologist Jean-Paul Bozonnet examines in his work the rise of the "great ecological narrative" that varies in degrees and intensities but shares a similar ecocentric worldview.[6] In this narrative that Bozonnet calls "ecocentrism," the Earth becomes the main protagonist of the story, replacing man in the "great Promethean narrative" that accompanied the birth of modern culture, and replacing God in the "great theocentric narrative" that preceded it. The adoption of these ecological values by the larger public happens at varying degrees, but its breadth is large, reaching "almost the entirety of the population of Western countries and an important part of the population in developing countries."[7]

This description does not apply as well, however, to the United States, where opinions vary widely. This is well documented in regards to climate-change issues,[8] for example. The politicization of climate sciences and the influence of climate skeptics with Republican backing have eroded the sense of human responsibility regarding climate change since 2008, as well as the belief in the consequences of climate change. Among Republican Party supporters, only 30 percent are now willing to admit the fact and effects of climate change, compared to 50 percent in 2008. Awareness among Democratic Party supporters, on the other hand, has remained fairly stable at around 70 percent over the last few years.[9] The United States is also home to the largest number of climate skeptics: 70 percent of the population believe that climate change is a reality, which is the lowest percentage recorded in countries around the world.[10] However, 70 percent is still a high proportion. And around 80 percent of the American population are concerned about the possible effects that environmental changes might have on their health.[11] This proportion is quite similar to that found in other countries, thus showing that even in the most skeptical country, environmental awareness is still important.

In France, Bozonnet studied the depth and nature of this new ecological narrative among the French population, using the paradigm test created by Dunlap et al.[12] Bozonnet found that 75 percent to 95 percent of the French population approved proposals moving in the direction of ecocentrism. "Pure" ecocentrics represented about 18 percent of the population, vs. 14.5 percent and 12 percent for pure Prometheans and pure theocentrics respectively.[13] These studies, although they apply to the French population, are in accordance with other sociological analyses, such as that of "cultural creatives," who represent 25 percent of the population in industrialized countries fairly consistently.[14] Bozonnet's studies are also in accordance with other analyses from influential thinkers such as Jeremy Rifkin, who asserts that the emergence of a "biospheric awareness" is now superseding religious ways of thinking, most notably monotheist and patriarchal;[15] and Bernard Lietaer, an economist specializing in currencies, who sees in the reemergence of local currencies over the last few decades around the world a resurgence of the goddess-mother archetype in the collective unconscious, whose correlative can be found in the High Middle Ages in central Europe as well as in Antiquity, and even perhaps in prehistoric societies.[16] Likewise, many international studies are showing that environmental concerns are not the privilege of wealthy populations in industrialized countries, and that they are instead also present in less wealthy countries. The free and wide distribution of the movie *Home*,[17] which is estimated to have been seen by 600 million people since its release in 2009, is an experiment that moves in a similar direction. This movie underlines the ecological, social, and economic shortfalls of the Promethean system that has dominated the past two centuries, while reminding its viewers of the fundamental link that exists between humanity, the living world, and the Earth. The film has been translated into many languages, including Central African, Latin American, and Asian dialects.

Finally, the past few decades have seen many examples of strong commitments from emerging and poor countries or from multiethnic communities in industrialized countries that invoke "Mother Nature," such as the Cochabamba Declaration of 2010, signed by 142 countries, and the Principles of Environmental Justice produced by leaders at the First People of Color Environmental Leadership Summit in Washington, DC, in 1991.

Even if these studies do not all use the ecological paradigm test at the global level, we can nevertheless put forward the hypothesis, given the homogeneity of other parameters, that the ecocentric paradigm is growing around the world, and that it is on its way to obtaining a consensual dimension.

## THE ROLE OF PERCEPTION IN GLOBAL ECOLOGICAL AWARENESS: AN AESTHETIC DIMENSION

According to Bozonnet, the consensual basis of ecocentric values attributes the status of myth to these values.[18] What is a myth? It structures our perception of and our relationship to the cosmos, society, and our own individuality. It has its own symbols and representations; it has its own aesthetics. Bozonnet defines this new myth, this "great ecological narrative," as autonomous, having little to do with the actual acceptance of threatening risks.[19] As such, the adherence to this narrative does not necessarily appear most prominently in the populations exposed to the most risk. Yet, the rise of this myth is contemporary to the birth of the ecological crisis and to the increased destruction of ecological equilibrium (ecosystems, climate, etc.)

To explain this apparent contradiction, I propose that aesthetics play an important role in the adherence to the ecocentric myth. Understanding the origin of risk necessitates a more global framework that allows for a structuring of perception. To define "aesthetics," I will use the definition of "aesthetic commitment" proposed by Nathalie Blanc, who works on this question from the perspective of environmental awareness:[20]

> We are not referring to a specialized field of studies like the arts, nor to a philosophy of the beautiful, nor to a theory about refined taste. We are referring to an active knowledge of one's surroundings, which is not limited to the arts nor to other cultural monuments . . . aesthetic experience is a way of inscribing the environment in oneself, and no longer making it an object of passive and disengaged contemplation.

This collective sense of engagement contributed greatly to the birth of the first international ecological organizations, such as the IUCN (International Union for the Conservation of Nature), created in 1948, and the WWF (World Wildlife Fund), created in 1961, that quickly became household names. Furthering this sense of belonging, in 1966, NASA released the first picture of the Earth from space. For the first time, we could see this planet that was ours: an Earthlight, an Earthball, lost in vast space, a single planet, finite, so small in the universe yet so big under our feet. Then came color photography, and we saw the Earth as blue (1972). This image, known as the "Blue Marble," aroused feelings of love, joy, and pride (see figure 1). Why have these pictures mattered so much, when they show us what we have known since the Copernican revolution that happened centuries ago? These images have added aesthetic experience to knowledge: we suddenly felt what we had only known intellectually. We came together with it—that is, we integrated it, we became aware of it. Better yet, these images of our planet were portraits: the Earth, suddenly, had a face. Personified, it sparked a stirring of empathy.[21]

Published by American biologist Rachel Carson in 1962, *Silent Spring* also contributed to this sense of empathy. For the first time, the impact of pesticide use on the environment was brought to public awareness. The book's reception was tremendous in the United States. Less so in Europe at the time, the "silent spring" is now part of the collective imaginary. It seeded a doubt in many minds about the harmlessness of these products and the consequences of their usage on our lives: one day, we might not hear the birds singing anymore. [22]

> There was a strange stillness. The birds, for example—where had they gone? Many people spoke of them, puzzled and disturbed. The feeding stations in the backyards were deserted. The few birds seen anywhere were moribund; they trembled violently and could not fly. It was a spring without voices . . . Future generations are unlikely to condone our lack of prudent concern for the integrity of the natural world that supports all life.

These words from the first few pages of *Silent Spring* illustrate that Rachel Carson's impact arises from the combination of scientific argumentation and vivid description of our sensory and aesthetic experience of the environment around us.

As ecosystems have become increasingly degraded, Rachel Carson's intuitions have been confirmed, resonating louder in the public consciousness. In 2005, over one thousand biodiversity specialists from all over the world released the first global report on the state of the planet's ecosystems, the *Millennium Assessment.* From 1950 to 2005, two-thirds of all

***Figure 1.*** THE EARTH SEEN FROM APOLLO 17. THIS PICTURE, TAKEN BY THE APOLLO 17 CREW IN 1972, BECAME KNOWN AS THE "BLUE MARBLE," ALSO THE NAME OF A 2012 SERIES OF HIGH-RESOLUTION DATA SETS FROM NASA. (SOURCE: NASA, HTTP://WWW.NASA.GOV)

ecosystem services—that is, air and water purification; soil fertilization; the provision of material, energy, and food; as well as the spiritual value of landscapes—have been degraded. In a little over fifty years, humanity has modified the Earth more than in its previous two thousand years of history. This report, which has been used in many texts about the environment (books, media, movies) targeting the wider public, has framed in a global context what most of us have now experienced at the individual level.[23] No matter where we live, we all have landscapes, rivers, beaches, or forests that marked our childhood but that we no longer recognize, or that have vanished. We have seen what seemed eternal, part of an unchanging background setting, modified during our short lifetime; we are left with a wound, as these places were part of our identity. The painful aesthetic experience of disappearing places combined with the knowledge of the global nature of this phenomenon produces a realization and an awareness that are both intimate and collective: what I am experiencing is similar to the experiences of all of humanity.

Different world summits and international conferences have emphasized at a political level our dependency on the global ecosystem. The haste with which world leaders and governments, whose actions are broadcast in every household via television and newspaper, have come to the Earth's aid has raised to a tangible and collective level the situation so well summarized by Michel Serres in his book *The Natural Contract*: "Through our mastery, we have become so much and so little masters of the Earth that it once again threatens to master us in turn. Through it, with it, and in it, we share one temporal destiny."[24]

In 1992, 120 world leaders and governments were present at the Rio Earth Summit.[25] This number has risen this high only one other time, at the climate-change conference at Copenhagen in 2009, where 130 leaders were present, again revealing the global scope of ecological awareness. Never before in the history of the UN have so many governments and world leaders been reunited.[26] This fact was significant: it meant that, at least symbolically, ecological issues were becoming as important as human issues, from which they could no longer be separated. The images broadcast on television not only provided important information, but also constituted an aesthetic experience, modifying the perception of our collective relationship to an Earth that is becoming, as Bozonnet puts it, the most important subject of human history.

## WHAT I DO TO THE LIVING, I DO TO MYSELF

Since the year 2000, another phenomenon has been growing, contributing both aesthetically and informatively to our increased ecological awareness of the fact that we are not only human beings dependent on a global environment, but that we are also living beings among other living beings. In my view, this step is the most crucial and the most consequential towards achieving an ecocentric awareness that views the human world and the living world as inseparable. Cancer has become a part of our families, as well as reproductive difficulties. Higher life expectancy explains only in part the increase in the number of cancers, which now affect one in every three women and one in every two men in industrialized countries. We know that the main cause of cancer is environmental, linked notably to synthetic compounds created in the last few years. Studies carried out on human populations across the world show the scope of the problem. To cite only one example among thousands, studies have revealed the presence of toxic molecules in the mother's umbilical cord at the moment of the baby's birth. Human babies are now born polluted, because of molecules that accumulate in their mothers' bodies and that are transmitted to them through a placental barrier unfit to block these substances.[27]

When the public learns that honeybees are dying en masse from multiple environmental factors, or that fish have cancer or are abnormally changing gender, all of this information begins to link together: the honeybees, the fishes, the uncle or the aunt in chemotherapy, the niece going through "such an early puberty," or the friends who started in vitro fertilization create a whole. Opinion polls show that nowadays 80 percent of the world population's main concern about the evolution of the planet is first and foremost related to their health and the health of their loved ones:[28] the ecological crisis deeply affects our hearts and our bodies.

Through this collective experience, we are slowly recognizing that what we are doing to the living world, we are doing to ourselves. We are all connected. The living world is not only the landscape surrounding us, it is in us. The aesthetic experience truly reveals itself to be "a way of inscribing the environment in oneself, and no longer making it an object of passive and disengaged contemplation."[29]

During this evolution, a slow shift has occurred from a vision of Nature as Other to subdue, to a vision of Nature as Other to protect and on which we depend, and finally, to a vision of Nature that loses its conceptual edges, as much outside as inside us: the Other that is Nature is now within ourselves. It has been interiorized. Over the course of these progressive shifts, we can see the three aspects of a great mythical narrative: cosmogonic, societal, and individual.

## THE ECOLOGICAL QUESTION: GOING BEYOND ZOOCENTRIC THINKING

The ecological crisis is causing a dramatic aesthetic and conceptual change. In terms of the aesthetic, this crisis triggers changes in our perception of the environment, both local and global, including at the level of our own body. We are discovering, in a tangible way, that we are linked to equilibriums that go beyond ourselves, that traverse the living world. This is also a conceptual change, because John Lovelock's hypothesis—that the Earth is home to a "cybernetic system with homeostatic tendencies"[30] produced by the interactions of all living organisms in a mineral, organic, and informational energy flux—is continually confirmed by present upheavals and the advancement of knowledge. The vast majority of publications on the current crisis and its causes refer to disruptions of external global equilibriums, such as climate and ecosystem resilience, and to disruptions of internal local equilibriums, such as immune or nervous systems, or cell reproduction (whose malfunction is at the origin of cancers). These disruptions create a new situation in the Western world, where we realize that we are living beings, an integral part of a living system upon which we have a profound effect.

This shift in the collective imaginary about our place within the living world can be related to the animal question that has been the subject of a long, rich philosophical tradition in the West. This tradition examines our place in, and our responsibility to, the animal kingdom to which we belong; the consequences of this way of seeing on how we perceive ourselves; and what this explains in terms of our way of behaving in the human world. The ecological question, as I have outlined it up until now, as born from the ecological crisis and therefore a much more recent issue, asks similar questions expanded to include the whole of the living world.

Just as with the animal question, the ecological question considers the human being as both object and subject of its condition of being a living being. But unlike the animal question, the ecological question holds that between the condition of being a subject and that of being an object, there exists a whole set of complex mechanisms of the living world. This complex system stands between our status as object and subject, and echoes Michel Serres's assertion about our new epistemological condition: "we depend on that which depends on us."[31] We are elements of a global system, made up of physical, chemical, and biological interactions that act and react in the short and long term, an incredible complexity that extends from the infinitely small to the infinitely large. We depend on equilibriums emerging from an intricate fabric of connections of which we are only beginning to understand the complexity.

The animal question and the ecological question do not, therefore, consider the same scales. The ecological question has a cosmogonic dimension. It goes beyond the animal kingdom and situates us in a more global and complex system.

By pushing us to think about a larger whole that goes beyond the animal kingdom, the ecological question situates the human being within a very heterogeneous group of life forms that have traditionally been divided into four kingdoms: bacteria, mushroom, plant, and animal. Because of their irreducible difference, these life forms require us to rethink the concept of the Other. Having caused extreme disruptions that affect the entirety of the living world, we have a responsibility to conceptualize and understand this world in such a way that we see our actions in a world a large part of which lives according to different temporal and spatial realities.

The great botanist Francis Hallé's way of thinking about plant life seems to me to be helpful in this endeavor. Hallé's work is all the more valuable because he brings to his writing not only a botanist's precision, but also the poetry and humility of a writer who addresses the general public. As he explains about plant life in his *Éloge de la plante* and his *Plaidoyer pour l'arbre*, we are faced with an "absolute otherness":[32]

> Biology is tailored for humans and animals, while the plant must repeatedly visit the bed of Procrustes. Fundamental concepts such as the individual, the genome, sexuality, species or evolution need to be converted or transformed in order to arrive at an objective biology, rid of zoocentrism and anthropomorphism.[33]

The ecological question forces us to think beyond our human condition and also beyond our animal condition so that we begin to think about our condition of being living beings. It seems to me that the ecological crisis is teaching us three fundamental lessons. First, it teaches us that we cannot escape our condition of being living beings. If we mistreat the living world, we are hurting ourselves. Second, as a traumatic experience for the West, the ecological crisis challenges our faith in science, our technologies, and our vision of human progress. Indeed, the ecological crisis teaches us that we are extremely ignorant about the conditions of life and the functioning of the living world itself. It is surprising that most of our knowledge about ecosystem services, and more tellingly our gratitude for these services, has happened largely after their disappearance. We are discovering that life is not the result of simple metabolic relationships within an organism; it is a set of interactions between multiple species, undoing our notions of identity and individuality, and our commonly held view of the living and the nonliving.[34] Third, the ecological crisis shows us, as Francis Hallé explains, that far from having an objective perception of the living world, we gather our knowledge about this world first and foremost through a thought process that is both animal and human. Even when we move away from animal observation to study plant life, we are limited by zoomorphic thinking, which constructs itself through zoomorphic language: "Our language itself is ill-adapted to plants, which are not knowing, do not use, have no needs, nor projects, nor goals. We are using an animal related language which lends itself poorly to the telling of a plant truth."[35]

The ecological question raises the issue of decentering with regards to our animality. Not only do we not have the traditions and customs for conceptualizing our place and responsibilities in the richly diverse living world, but we are also very limited by our zoomorphic language. Does what Francis Hallé note about the sciences not also hold for the humanities?

What would a philosophy, an economy, and a sociology be like if imagined by thinking in a vegetal mode or more generally in an ecological mode?

From a philosophical standpoint, the ecological question and the animal question come together on one fundamental point: we are a species of radical departure, which is to say, a species that brings new dimensions to the living world, such as the connection between self-consciousness and language and the ability to act on the world in a significant way.[36] This particular power imposes on us a particular responsibility. Our arrival as a species into the animal kingdom represents a break, and has led to many disruptions that have affected the conditions relating to the life and development of many other animal species, through domestication and breeding, fracturing of habitats and migration routes, technologization of animal bodies through genetic selection and industrialized farming. As an indirect result, these technologies used on animals are available to be used on humans. The only limits are ethical and moral considerations whose definitions remain unclear and debated. The philosophical tradition that has examined for many centuries the question of the differences and similarities between humans and animals has established some key milestones: What does it mean that we have a capacity for reflexive thinking, giving us an unprecedented power over all life forms, including ourselves? What does it mean to trace boundaries between ourselves and other species? What does it mean for our societies and each of us as individuals to learn that we are not in an ivory tower, as we had previously thought?

These are some of the questions raised by the ecological question, and for which the philosophical tradition of the animal question provides a large body of literature that can be revisited under the wider lens of the ecological question.

## ECOLOGICAL AWARENESS: RETHINKING WORLDVIEWS AND TECHNOLOGIES

The ecological crisis could be compared to the tip of the iceberg, while the submerged part, much deeper and wider, would be a cultural crisis,[37] which is revealed through the artifacts of our thinking that have led us to consider humans as separate from the rest of the living world, and to underrate the relationships essential to our shared viability. Our ecosystems are no longer a landscape where our activities occur, or a pool of resources; they have become a dynamic network of which we are a part as living beings, but most especially as thinking beings, the only species able to conceptualize this network because of our capacity for reflexive thought. This vision of ourselves as living beings inscribed in the network of the living world generates an unprecedented opportunity to renew Western ways of thinking within both the hard sciences and the humanities.

## THE RENEWAL OF ECONOMIC AND APPLIED THINKING

Since the 1990s, new concepts have begun appearing in the hard sciences and in the fields of economy and production. For example, William McDonough and Michael Braungart have proposed the principle of "cradle to cradle," according to which human activities, to be viable, must follow the circular model of the living world where one group's waste represents another group's resource. More generally, biomimicry, a principle first theorized by Janine Benyus in 1998,[38] conceives of the living world as a vast community of intelligence accumulated over three and a half billion years, whose areas of innovation have as much to do with human organization as with habitat, locomotion, and production, all leading progressively to an economy of energy, matter, and increased biodegradability.

The rapid development of our knowledge about ecosystems—a knowledge that was almost nonexistent before we became interested in the benefits of ecosystem services—is revolutionizing the way we think about plant life. Plants, whose main characteristic is immobility, have developed highly sophisticated strategies in order to conquer territories, satisfy their energetic and dietary needs, and degrade their waste. As Emmanuelle Grundmann explains, "the tree doesn't adapt to its environment; it creates it."[39] We are even discovering that the root apex is probably the site of a nervous signal that acts on the root exudations in a manner similar to that of neurons and synaptic activity.[40] And we already know that aerial communication between species of plants and/or animals is extremely complex and efficient. Finally, sociological, psychological, and neurophysiological studies show that we maintain relationships with plants without even knowing it because their presence and diversity regulate our stress and contribute to our physical and mental regeneration.[41]

## A NEW PHILOSOPHY OF THE LIVING WORLD

We can see a new philosophy of the relationship between humans and the rest of the living world beginning to emerge. It is no longer a question of observing nature in order to better subdue it, as Francis Bacon asserted in the sixteenth and seventeenth centuries; it is a question of observing nature to better reintegrate it.

These recent developments in our knowledge about the living world give rise to another dimension: a sense of wonder. Our belonging to the living world is not an accident of which we are victims, and whose impact and scope we measure in order to continue to exist as living beings, physiologically, physically, psychologically, and socially healthy. Knowledge of the living world can lead to a sense of pride in belonging to this world whose intelligence at various levels is nothing less than mind-boggling.

A whole field of possibilities is opening up. For over four decades now, we have been witnessing all around the world economic initiatives responding to the ecological crisis. They allow us to imagine an economic order where human activities align with the restoration of ecosystems—that is to say, an economy in symbiosis with the living world, in which the activity of one part benefits the others. In urban design, we now know how to regulate microclimates, to insulate buildings by integrating green roofs, to filter water by developing humid ecosystems at the foot of buildings, and to produce high food yield on very small areas. What other species can re-create, on less than 100 square meters, ecosystems such as a steppe on the roof of a building, humid zones and forest gardens at its feet, and living space for humans between the two? Possibly only trees can bring together in such a small area such a variety and wealth of species.

What we can see here is the power of going beyond thinking about animals who do not create their own matter and for whom the environment provides stocks on which they can feed. Indeed, over the last three and a half billion years, during which photosynthesis has been active on Earth, the living world has continually proven that it is perpetually growing. If one thinks ecologically, the living world is a whole in which the human is inscribed and to which we can contribute through our capacity for observation, conceptualization, and organization.

These discoveries have implications for the common good and therefore have an important political dimension. They raise questions about our societies because, with respect to structures and ultimate needs, cities have much in common with ecosystems; the latter only rarely assume hierarchical and centralized forms of organization and exchange.

This realization of being living beings, born from the awareness of our own fragility and dependence and from our fear, is currently being transformed into a positive awakening. It has the power to change things. It has philosophical leverage that initiates debate about a necessary redefinition of our idea of the law in order to make legal subjects of this "third nature" (Serres), these "non-humans" (Latour),[42] or this living world to which we, human beings, also belong. It has anthropological leverage for rereading the history of our societies, following the example of Frans de Waal after his discoveries of empathy and collaboration in the animal world—discoveries that motivated other influential thinkers such as Jeremy Rifkin.[43] This new awareness is an inspiration for revisiting the history of our relationship to the living world, as the growing movement of ecocriticism is doing in Anglo-Saxon academic circles using art and literature. Finally, it also has political leverage, and it is interesting to note that the World Social Forum, which began in Porto Alegre (Brazil) in 2001 around questions of democracy and poverty, is now becoming a forum on "biocivilization."

## "THE HUMAN IS THE LIVING WORLD BECOMING CONSCIOUS OF ITSELF":[44] THE AESTHETICS OF THINKING ECOLOGICALLY

Communicating the advances in our knowledge to as many people as possible does not fulfill one single goal, be it scientific, practical, or intellectual. These advances renew our imagination and open up pathways in order to get us out of the ecological dead-end we presently face, and reorient our decisions about our modes of production and consumption, our ways of developing land and responding to our needs.

If, as I believe, aesthetics has played a fundamental role in the growing awareness of global dangers, it is also key to communicating the progressive side of an ecological awareness that recognizes both the specific intelligence of human beings and the intelligence of the living world to which we belong. Ecological awareness brings about a shift in our perception of belonging: we see it from above, as if from a step up. This awareness renews our imagination because it gives rise to a different vision of things with which we have long been familiar. It brings us into contact with a newly perceived environment in which a complex living world finds once again its place. The aesthetic phenomenon happens when we come into contact with the representation of an object that modifies the perception we had of this object. When information combines with feeling, a new cognitive connection is created, infused with emotion whose intensity engraves the perception that much deeper in our minds.

We know that aesthetics can play an important role when communities take action to restore ecosystems where the human world has taken over.[45] I believe that the role of feeling is also crucial to thinking ecologically, that is to say, a way of thinking that perceives the human, who has his own specific intelligence, as part of a dynamic network of links in the living world that has its own forms of intelligence. In my view, thinking ecologically is in essence an aesthetic transformation: it leads us to another *perception* of ourselves, another *perception* of the living world, and another *perception* of the links between us and the living world. It seems to me absolutely necessary to take into consideration the aesthetic dimension of thinking ecologically, and to deepen our understanding of its role while reflecting on these new models and communicating their usefulness.

A movement has started, opening new pathways for the West and all of humanity, leading towards a new vision of the living world and the place of humans in this living world. But we cannot wait for this to happen on its own. The ecological clock is ticking: the links that are holding together the global ecosystem could break at any time. This could happen in three decades, or two, or one. Maybe it is already too late. The faster this movement grows, the more possibilities we will have to solve this ecological crisis. In this chain reaction of collective thought, the recognition of the aesthetic dimension of thinking ecologically is—I am convinced of this—an essential catalyst.

## NOTES

1. The "living world" is a translation of "le vivant," a term in French that is difficult to translate into English. "Le vivant" is different from the concept of "nature" that often follows the paradigm of a separation between the human and the living world; see Isabelle Delannoy, "Sciences, humanités et écologie—trouver les nouvelles passerelles et les nouveaux mots pour entreprendre un nouveau monde," *French Literary Studies* 39 (2012): 139–154.
2. See Jeremy Rifkin, *La troisième révolution industrielle: Comment le pouvoir latéral va transformer l'énergie, l'économie et le monde*, trans. Françoise and Paul Chemla (Paris: Les Liens qui Libèrent, 2012).
3. See Jean-Paul Bozonnet, "Les métamorphoses du grand récit écologiste et son appropriation par la société civile," *Revue d'Allemagne et des pays de langue allemande* 9 (2007): 311–342.
4. Postdam Institute for Climate Impact Research and Climate Analytics, Report for the World Bank, *Turn Down the Heat: Why a 4°C Warmer World Must Be Avoided*, November 2012, http://climatechange.worldbank.org/sites/default/files/Turn_Down_the_heat_Why_a_4_degree_centrigrade_warmer_world_must_be_avoided.pdf.
5. It would be wrong to attribute these models to all of humanity; they are a product of Western civilization, whose values, worldview, and technology have largely spread around the world, since colonial times and the industrial revolution, with rapid acceleration since the 1950s.
6. Jean-Paul Bozonnet, "Les métamorphoses."
7. Jean-Paul Bozonnet, "Les métamorphoses," 330–331.
8. There are more opinion surveys on climate change than any other issue.
9. Nomadéis/K-Minos/Semiocast, Report for the Centre d'analyse stratégique, *La perception internationale du discours scientifique sur la menace climatique par le grand public dans six pays: Afrique du Sud, Brésil, Chine, États-Unis, France, Inde*, October 2012, http://www.strategie.gouv.fr/system/files/rapport_etude_cas_nomadeis_annexes.pdf.
10. AXA-Ipsos, *La perception individuelle des risques climatiques*, 2012, http://www.axa.com/lib/axa/uploads/cahiersaxa/Etude-AXA-Ipsos_risques-climatiques.pdf.

11. Edelman "goodpurpose" study, 2010 and 2012. Surveys of 8,000 people, 18 years old and above, in 16 different countries: Canada, Brazil, USA, United Kingdom, France, Netherlands, Belgium, Singapore, China, India, Malaysia, Japan, Germany, Indonesia, United Arab Emirates, Italy, http://purpose.edelman.com/.
12. Riley Dunlap, Kent Van Liere, Angela Mertig, and Robert Emmet Jones, "Measuring Endorsement of the New Ecological Paradigm: A Revisited NEP Scale," *Journal of Social Issues* 56 (2000): 425–442.
13. Jean-Paul Bozonnet, *L'imaginaire écocentrique: Un récit postmoderne contre les institutions*, paper delivered at the Journée d'étude du C2SO et du CEDRATS: "Interaction Homme–Nature, Entre imaginaire romantique et gestion politique de la nature, quelles sont les dimensions pertinentes pour penser, analyser et agir sur les questions environnementales?," Lyon, April 23–24, 2010.
14. Paul Ray and Sherry Anderson, *The Cultural Creatives: How 50 Million People Are Changing the World* (New York: Harmony Books, 2000).
15. Jeremy Rifkin, *Une nouvelle conscience pour un monde en crise: Vers une civilisation de l'empathie*, trans. Françoise and Paul Chemla (Paris: Éd. Les Liens qui libèrent, 2011).
16. Bernard Lietaer, *Au cœur de la monnaie: Systèmes monétaires, inconscient collectif, archetypes et tabous*, trans. Michel Icks (N.p.: Yves Michel, 2011).
17. *Home* is a documentary directed by Yann Arthus-Bertrand, produced by Elzevir Films and Europacorp, available for free on the Internet, in movie theaters, on television, and shown for the first time on June 5, 2009.
18. Bozonnet, "Les métamorphoses."
19. Bozonnet, "Les métamorphoses," 312–317.
20. Nathalie Blanc, "Éthique et esthétique de l'environnement," *EspacesTemps.net*, January 2008, http://www.espacestemps.net/en/articles/thique-et-esthetique-de-lrsquoenvironnement-en/.
21. It should be noted that this first photograph of the Earth was not calling into question the dominant Promethean myth. Quite the opposite: it was reinforcing it. It was new evidence of the superiority of man, who, thanks to an amazing technological savoir-faire, was capable of escaping gravity to observe the Earth from space.
22. Rachel Carson, *Silent Spring*, 40th Anniversary Edition (New York: Houghton Mifflin, 2002), 2, 13.
23. In 2012, the AXA-Ipsos study "La perception," carried out on three continents and thirteen countries, developing and developed, showed that 73 percent of the people surveyed declared having witnessed and individually experienced climate change; 30 percent stated that climate change was already affecting their personal comfort level.
24. Michel Serres, *The Natural Contract*, trans. Elizabeth MacArthur and William Paulson (Ann Arbor: University of Michigan Press, 1995), 33–34.
25. Matthieu Baudin, *Le développement durable: Nouvelle idéologie pour le XXIe siècle?* (Paris: L'Harmattan, 2009).
26. Janos Pasztor, UN director of the Climate Change Support Team, conference address at the Fresh Air Center, Copenhagen, December 17, 2009.
27. Jacqueline Schuiling and Wytze van der Naald, *Toxiques en heritage*, trans. Alex Freiszmuth and Yannick Vicaire (Amsterdam: Greenpeace International, 2005); Maria José Lopez Espinosa, "Maternal-child Exposure via the Placenta to Environmental Chemical Substances with Hormonal Activity," PhD diss., University of Grenada, Spain, 2007; ASEF, *Femmes enceintes et produits chimiques, la synthèse de l'ASEF*, 2012, http://www.asef-asso.fr/mon-enfant/ma-grossesse/1255-femmes-enceintes-et-produits-chimiques-la-synthese-de-1-asef.
28. Edelman, "Good Purpose."

29. Blanc, “Éthique et esthétique.”
30. Lynn Margulis, “Gaïa,” in *Écologie politique, cosmos, communautés, milieux*, ed. Émilie Hache (Paris: Amsterdam, 2012), 256.
31. Michel Serres, “Retour au contrat naturel,” *1000 Days of Theory, C-theory.net,* May 2006, http://www.ctheory.net/articles.aspx?id=516.
32. Francis Hallé, *Plaidoyer pour l'arbre* (Paris: Éd. Actes Sud, 2005), 167.
33. Francis Hallé, *éloge de la plante: Pour une nouvelle biologie* (Paris: Seuil, 1999), 322.
34. Recent discoveries about the immune system and bacterial equilibrium within our bodies show that the notion of the individual is becoming blurry from a biological perspective. Thus we are discovering that the bacteria in only one human body outnumber human cells, that their combined genomes are greater than the genome of the individual, and that in one body they are even greater than in the combined genome of all of humanity. Their activity affects the equilibrium of our immune system, but also that of our digestive, nervous, and hormonal systems. These bacteria influence our moods and our physical appearance. It has been suggested that bacterial imbalances due to the intake of antibiotics modify the flora that regulates appetite and thus could be partially responsible for the increase in obesity and overweight rates. We are also discovering that genetic transmission is not only vertical through a line of ancestors, but also horizontal through genes transmitted by bacterial and viral infections, which can also modify a species; 8 percent of human genes could be the result of such horizontal transmissions, a proportion that increases to 70 percent for corn.
35. Hallé, *Éloge*, 325.
36. This break is now attributed to the combination of three particular characteristics: an extremely important neocortex, more complex than that of other mammals; the Broca language center and a long enough larynx to allow for articulation; and the hand.
37. This image is borrowed from Francesco Lozano Winterhader, speaker and consultant in holistic sustainability, from a paper given at the Goodplanet Foundation, Paris, October 2012.
38. Janine Benyus, *Biomimicry: Innovation Inspired by Nature* (New York: William Morrow, 1997).
39. Emmanuelle Grundmann, *Ces forêts qu'on assassine* (Paris: Calmann-Levy, 2007), 57.
40. František Baluška, Stefano Mancuso, Dieter Volkmann, Peter Barlow, “Root Apices as Plant Command Centres: The Unique ‘Brain-like’ Status of the Root Apex Transition Zone,” *Biologia* 59 (2004): 1–13.
41. Plante et Cité, “Bienfaits du végétal en ville sur le bien-être et la santé humaine,” October 2009, http://vegepolys.eu/media/rpc__n_special_bienfaits_nature_en_ville__015515700_1226_18122009.pdf.
42. Bruno Latour, *Nous n'avons jamais été modernes: Essai d'anthropologie symétrique* (Paris: La Découverte, 1991).
43. Jeremy Rifkin, *The Empathic Civilization: The Race to Global Consciousness in a World in Crisis* (New York: Penguin, 2009).
44. The title echoes Élysée Reclus's epigram: “Man is Nature becoming conscious of herself” (“l'homme est la nature prenant conscience d'elle-même”).
45. Blanc, “Éthique et esthétique.”

# The Greenway
## *A Study of Shared Animal/Human Mobility*

NATHALIE BLANC

*Translated by* CHRISTINA SUTTON AND LOUISA MACKENZIE

Since 2007, following the so-called *Grenelle 2 de l'environnement* (the second Grenelle "round table" on the environment), greenways have achieved stunning success in France.[1] Integral to the rapidly growing science of landscape ecology, greenways form a mesh of planted spaces within urban environments. One of their main functions is the restoration of biodiversity (of both plant and animal species) that in turn provides more genetic resources and a greater variety of ecosystems. The creation and maintenance of greenways is causing our conception of metropolitan spaces to evolve, and forcing us to rethink how we represent and understand animals within the cityscape. What must now be considered is how to think not only about the presence of animals, but also about the unusual cohabitation of humans and animals within territories that are usually understood as being exclusively for humans. In light of biodiversity—a key word, used by scientists and the media alike, that situates animals within an ensemble of living beings essential to the survival of humans—how should we rework the stories we tell about the multiple encounters between humans, individual animals, and groups of species (cats, dogs, etc.)?[2] How will future encounters, both welcome and unwelcome, between humans and animals in urban environments make us reconsider our ethical relations with animals and, more specifically, to revise our behavior towards them in order to maintain or achieve a full and happy life?[3]

The essential question is and remains: what is the relationship between a wanted and an unwanted nature? Contrasting perceptions of both reveal the complexity of this question. The former is often considered a tool in urban design, and involves several plant and animal species seen as purely decorative, while the latter consists of diverse species that are labeled nuisances by social representations and practices, and designated as invasive by scientific specialists. Of course, many types of "nature" exist: one that is desired, controlled, and produced; one that is a result of the city/nature interface (cockroaches, pollution, etc.); and one that escapes human control because of its exceptionality or quantity (floods, earthquakes, swarms, etc.). This complication of opposites is a defining element of the method I have developed through my research. This method teases out the concealed and revealed dimensions of nature in the city in order to better understand the issues at the heart of the "ecological" relationship more generally. It also enables further reflection about social and geographical configurations by thinking in terms of the *reciprocal* agency of human beings and elements of the natural and constructed environment.

In what follows, the first section will examine the way in which the opposition between wanted and unwanted animals is structured, and will present some conclusions on what characterizes an urban environment. The second section will explore how these ideas inform a relationship to the environment and more specifically to greenways (environmental policy tools that produce a "desired" natural environment as a way of combating the loss of biodiversity) that is in fact quite limiting.[4] Although numerous environmental policies draw on scientific studies that put forward a complex systemic vision of natural dynamics, these policies still fail to articulate these questions with an analysis of the place of the plants and animals within a given environment. This place, whether it be social, geographical, political, or aesthetic, is neglected and elided. It is particularly important to reflect upon the representations and practices of the animal in relation to the development of greenways, a reflection that will serve as the basis for the fourth and fifth sections of this article. For if representations and practices inform urban narratives and modes of cohabitation with animals, greenways themselves are constructed using images and maps, all fixed modes of representation. Despite this fixity, though, one of the central dimensions of the contemporary animal question from the perspective of environmental policy is, in my understanding, a personal relationship to the living world—a relationship that evolves in time and space according to specific modalities. I will thus briefly consider some ethical questions concerning the appropriate place of animals, and how we should interact with them.

## RESEARCH TRAJECTORY

Since the 1990s, my research has focused on nature in the urban environment, interdisciplinary studies involving colleagues in the biological and material sciences.[5] The urban environment is a place largely ignored in environmental research, which essentially focuses on rural or wild natural spaces. To explore the natural dimensions of urban landscapes, both their representational and material aspects, can be innovative and risky.[6] This is all the more true because the goal of this work is primarily to start from the place of the animal in order to understand the natural environment within the city. However, for many who live in the urban environment, nature within the city is all about plants. Furthermore, when considering animals, I was focused on two species in particular: an insect of tropical origin, the cockroach, entirely unwelcome in the urban environment; and the cat, a mammal that is either feral or domesticated, and often fed by groups dedicated to protecting them. The goal of my research was to compare representations and social practices concerning animals, with scientific knowledge relative to their presence in urban spaces (numbers, group dynamics, characteristics of adaptation, etc.) These case studies showed that cockroaches are not considered to be a part of urban nature: viewed neither as wild (because present in a human environment) nor domestic (because they avoid human contact), they are most often treated as pests, symptomatic of a dirty and polluted city, populated by the poor. This unwanted animal attests to the challenges, greater or lesser according to the social status of the occupants, faced by urban management approaches. On the contrary, the stray cat (sometimes called a "free" cat)[7]—whether feral or an escapee from the domestic environment, desirable when owned, pampered, and fed, and unwanted when its health or feline character becomes a source of anxiety—does represent nature (all the more so when less domesticated). This work thus highlights the notion of "control techniques" pertaining to different geographical spaces (for example, the space under the refrigerator can be considered a legitimate environment depending on who occupies it), and the ideas of "wanted" and "unwanted" natures. This research also illustrates the ways in which urban space is regulated to different degrees by a management system that determines nature's place and defines the human habitat. Finally, this project included the collection of various stories or narratives about urban animals that raised a series of questions: How do we narrate animal presences in the city? To which urban functions do these representations have relevance? Does each animal or species have its own particular narrative that functions within the urban space? How does time combine with space in the making of the urban animal in the present?

Jacques Derrida explains that the term "animal" (like that of "man" to some extent) designates one and only one being, subsuming all differences. Derrida employs the term "animot" to demonstrate the extent to which "animal" is only a word.[8] Throughout my research, singular encounters in different French towns with animals and with the people whose lives intersect with them have required me to contextualize the presence of animals. These particular encounters reveal the richness and extreme complexity of modes of human-animal cohabitation in the urban environment. In this way, modes of companionship are defined as much by animal natures as by human cultures made possible in specific spaces. Cockroaches, for example, are the product of urban management inasmuch as the modern constructed space is their ideal habitat: warm and humid with food and shelter readily available and abundant. At the same time, cohabitating with this species of insect represents a symbolic threat to the concept of a clean city while simultaneously affirming a truly "natural" environment for the cockroach, since tall city buildings serve as ecological niches to this animal of tropical origin. Those who look after stray cats are often characterized as pitiful, lonely women whose lives typify the problem of urban solitude; because they have no children, they supposedly have nothing better to do than look after animals. Having begun to work on the natural environment in the city and the relationship of cities to their natural environments, I have come to realize that nature is not just greenery, green spaces, or replanted landscapes in the city. Urban nature is also unwanted nature, the shadowy underside of the nature produced by urban spaces.

In other words, looking anew at the role of nature in the city requires us also to look at "unwanted," "uncontrolled" nature.[9] My research projects have included other animal species (rats, pigeons, etc.) and other elements of urban life, most notably vegetation. Since my recent research concerns biodiversity, I have been able to study the ways in which green city planning, intended to counter the loss of biodiversity in the context of urban policy, has translated into an increase in animal numbers and sightings. In what follows, I will focus my attention on the modalities of shared life, rather than simply on the status of the animal. What does it mean to live with animals in a space where bettering the conditions of human life is no longer synonymous with destroying the natural environment, in a space where each encounter, each interaction creates a connection with the animal that is revised and corrected in light of new issues and concerns?

My previous research has shown that our understanding of the place of animals in the city—an environment that had supposedly caused their disappearance—is complex and depends on social relations between humans. The presence of cockroaches in the projects or underprivileged suburbs, for example, reinforces the marginalization of those who inhabit these already socially stigmatized spaces. There is currently a movement that seeks to revalorize the presence of animals in cities for the sake of biodiversity. This shifts the focus from the concept of the animal, or animal nature, to concepts such as ecological landscape, biodiversity, and the living environment. However, it is unlikely that revalorizing that which is living in the urban environment will revalorize stigmatized neighborhoods in the eyes of those who live there. One can simply say that the underlying determinism of the methods used by professional urban planning to act on and produce urban environments has been under critical scrutiny for quite some time. As for animals, how we imagine them in future urban environments means reflecting on both the possible encounters with, and the reactions to, their presence. Since these animals will no longer be kept at a distance, we must begin to think in terms of co-presence. What sorts of fears, supposed, imagined, or harrowing, do different animal species produce? What pleasures might we derive from these encounters? Might the question of biodiversity then lead to new possibilities in the reappropriation of urban spaces? What kind of relationship between wanted and unwanted animals could be established on the basis of these new developments?

## GREENWAYS IN FRENCH CITIES

Could the question of biodiversity lead to new possibilities in the experience and uses of urban environments, and of wanted and unwanted natures? This notion seems to be at the heart of scientific studies on greenways. The underlying hypothesis is that urban biodiversity is a way of revalorizing the city in the eyes of its inhabitants, while also responding to environmental concerns about the loss of biodiversity. However, few studies are asking about the sort of nature generated by greenways, or about the instrumentalized role that living animal and plant populations will play in the resolution of the contemporary urban environmental dilemma. Recently, within the framework of a research project focusing on greenways in seven different French cities, I examined the actual and imagined place of animals in urban settings corresponding to these new developments.

Begun in 2009 and funded by the Agence Nationale de la Recherche (ANR), this research program—"Evaluation of Urban Greenways and Elaboration of Reference: An Infrastructure Balancing Esthetics and Ecology for a New Urbanity"[10]—adopted an interdisciplinary methodology that respected the complexity of the issues relating to the implementation of greenways. Approximately fifty researchers from ten different research teams established a project to evaluate greenways from cultural, sociological, geographical, and ecological perspectives. Debates were organized in three very different French urban conglomerations—Paris, Marseilles, and Strasbourg—in order to further understand the place of greenways in these three settings by asking what types of relationships to nature, to the animal, and to biodiversity were possible in different contexts. In addition to aiming to improve the quality of city life for humans, the environmental policies in these cities targeted plant and animal species, but only those considered "desirable." This raises the question about those "other" species, the ones deemed invasive or simply nuisances, a question that I am particularly interested in exploring further.

Paris, Marseilles, and Strasbourg each have development projects (subdivisions, greenbelts, etc.) under discussion and/or underway that promote continuous green spaces both as a way to benefit citizens, and as a way to bring about "natural" urban environments. These similarities are, however, contrasted by some important historical, sociological, and geographical differences. Paris, the nation's capital, is a city located within a larger urban zone exceeding 10 million inhabitants and extending to 2,845 square kilometers for the official urbanized area. The city center, proportionally quite small (105 square kilometers), is highly built up and characterized by high urban architectural and human density. With a population of 2.2 million *intra muros*, the human density is 20,980 inhabitants per square kilometer. Marseilles, France's second largest city, with a population of 851,000, is a metropolis with natural geographical boundaries. The borders include the Mediterranean Sea to the west; the hilly ranges of l'Estanque, l'Etoile, Saint-Cyr, and Marseilleveyre; and the mountains Puget and Garlaban covering the rest of the perimeter. With 3,538 inhabitants per square kilometer, the center is much less dense than Paris *intra muros*. The greater commune is, however, extremely expansive: at 240 square kilometers it is one of the largest in metropolitan France, a contributing factor in its lesser density. Finally, Strasbourg houses a third (272,000) of its 758,000 inhabitants on a communal territory of 78 square kilometers within the urban center, a density equivalent to that of Marseilles with 3,477 inhabitants per square kilometer. Strasbourg, the "European capital" that extends beyond the Rhine to connect with the German city of Kehl, is rightly proud of its contributions to building a sustainable urban territory, even if greenways as defined by the *Grenelle 2 de l'environnement* are still under discussion.

Eight different focus groups per city, each with six to nine members of the public working with two or three researchers, meant that a total of twenty-four groups with more than 160 city residents participated in the research project. The diversity of profiles among the focus groups and the individuals interviewed facilitated the matching of different environmental sensibilities and social and cultural representations of nature, with the environmental policies of each conglomeration and the representative urban plans. The moderators proposed a series of six questions. Drawing on a homogeneous corpus, our content analysis then classified and organized the discursive elements provided by individuals into communicative elements.

## INCREASING PLACES

Nature in urban areas most often means parks, because nature is understood primarily as vegetation. This vegetation is essentially a desired nature constructed in the spaces of parks, gardens, or balconies. However, animals (the fauna) are increasingly understood as part of the urban natural environment. This inclusivity demonstrates the changing place of urban animals in city dwellers' consciousness since the beginning of the twenty-first century, which may have to do with an overall increase in sensitivity to environmental issues.[11]

As for the research results, broad categories were cited (grasses, trees, animals, insects, birds, fish), while individual species were less likely to be mentioned, most notably in Paris. The species listed varied, of course, according to the city. Seagulls, wild boars, rabbits, and goats are typical of hilly, Mediterranean Marseilles. Strasbourg residents all mentioned storks, swans, and river rats (coypu). In Paris, seagulls (along the Seine), falcons (Notre-Dame), pigeons, and bees were mentioned. Domestic animals (dogs, cats, chickens, and ducks) were not considered a part of nature, while "urban" animals such as rats and pigeons were seen as reminders of the natural, linking the city with nature. Furthermore, walking a dog in a park or simply owning a companion animal was considered a means of getting closer to other living beings and also reconnecting with nature. Informant F (Paris, central-peripheral neighborhoods, environmentalist) said, "I had a dog, and just the act of owning a dog made me frequent parks more often. So, that actually brought me closer to nature."

During interviews, city dwellers concluded that "urban" animals belonged to a "false" urban nature. At the same time, these undesirable animals were associated with dirt, violence, and disorder. They symbolized the difficulties inherent in controlling urban nature. This "uncontrollable" nature took two forms: the unwanted urban animals, seen as unlikable although also unavoidable; and animals considered "wild," rarer and more valued, seen as part of real nature and not the city. Informant H (central Paris) said,

> If you go to the 18th district at the edge of the ring road, it's less beautiful: there are rats everywhere at night, even pretty rats. They're animals, yeah . . . Sometimes you also see squashed rats at Châtelet les Halles . . . Let's not kid ourselves; they are everywhere in Paris . . . I think there are a lot of animals, but you have to know how to find them. I think there may be around 2000 different animals in Paris. Insects, for example—in the park at Bercy, next to the bars, there are little insect houses, not a lot of people notice them.

Animals therefore have their own geographies, with certain species associated with certain places or territories. But these areas are often those that have been neglected by humans and become the margins of the urban environment. This is the case of the space under sinks, where cockroaches find refuge. Consequently, animals can still surprise us. Their mobility, their behavior, and even their needs reveal that which is unexpected in the urban environment. Does this way of living then constitute a form of liberty? It is, first and foremost, a source of untapped beauty, as numerous observations throughout the investigation attested: F (Marseilles, center, non-environmentalist) said, "Geckos, there's a bunch . . . There are some in the Panier neighborhood, they are everywhere . . . When I water my terrace garden, there are lots that come out from behind the pots, and it is very pretty." In such moments, the limits of the city become blurred, appearing more and more like a vanishing point. The surprise of seeing geckos might be appreciated, but it might also arouse fear of the sudden appearance of animals when they are least expected, in new and unwanted contexts. In short, no one wants to find a cockroach in their food.

But with regard to the pleasures that come from encounters with animals in the city, one of the most important ones—which justifies the idea of a desired biodiversity encouraging animal presence—is the way in which animals remind us of the sounds and smells of the city, of a specific urban beauty. The animal as perceived through the buzzing of a bee, the song of a bird, the sight of a deer, these are all things that make up the beauty of a city, a beauty that is closely related to the ecological sentiment of wanting to protect nature.[12] This is a beauty that plays to all five senses as well as to our sensory memory. Numerous projects relative to nature in the city are linked to the pleasure that one feels while observing the beauty of various animal and plant species. Witness this conversation between F and H (Strasbourg periphery-center, environmentalists). F: "[It's] pleasure. It's nicer to hear a magpie singing at 5 in the morning than to be stuck at a red light, I think it's about pleasure . . . Am I wrong?" H: "No, that's it, it's pleasure." F: "It also reminds you that you're not all alone in the world." H: "Yes, exactly, we're sharing." F: "We're all animals just like the others." H: "Yes, that's it." But without a doubt, if biodiversity is part of nature, animal biodiversity in the city is not always the most pleasant. F (Paris center, environmentalist) said: "Nature is a bit of everything. It is not always, or even often, for the greater well-being. Rats, for example, living freely in the sewers, profiting from the city like wild animals . . . That's not nature. It's an adapted animal, it's nature adapted to the city."

It now becomes important to see how we go from thinking in terms of "wanted" and "unwanted" nature to appreciating the beauty of nature that has adapted to the contingencies of city life. Wanted nature is also often the most domesticated, found on balconies, in window plantings and flowerpots, or in parks. It is a controlled and produced nature, instrumentalized as a kind of urban furniture. Yet for some years now, the trend of the "natural garden," or urban fallow land, has helped people, especially those concerned with environmental issues, to better appreciate the presence of spontaneous species in the city. Spaces and species that were once seen as the result of negligence or bad gardening are now qualified as beautiful. *Nature* is therefore also understood as being relatively free from human management practices, the *wild* as what is free from human activity. Nature that has adapted to the city includes species produced in urban environments, such as companion animals, but also spontaneous urban species. Some of these animal species are viewed less favorably because of how they have adapted to the urban environment: they are labeled either as nuisances (like the cockroach) or garbage species (like rats or pigeons). It is important to note that some species, such as the domestic pigeon, were once well regarded but are now considered very negatively. Produced, urban-adapted nature is now seen as city property, not as the beautiful nature of forests and wilderness, composed of noble animal species, largely mammals.[13] Furthermore, the animals that are more or less desirable in the city do not participate in the spectacle of urban nature. Could the introduction of these species using greenways play a role in valorizing a new kind of landscape, one in which animals would participate?

## MOVING TOWARDS ANIMAL LANDSCAPES?

Do animal landscapes exist? If landscapes are made up of vegetation, animals are but passing and decorative occupants in these urban images. Greenways, in this sense, constitute "regreened" spaces of reunification where animals and humans share territories. In Strasbourg, where greenways are already well established, and where animals and humans participate in a shared mobility, the testimonials are explicit. F (Strasbourg, peripheral-center, environmentalist) says: "Nature, it's greenery, but it's also animals. It's all that I see from my bike, going to and coming from work. For me, just being on my bike and surrounded by nature, a swan, a duck, well, I don't know, the birds, to me, all that clears my head. I feel well, I feel good. That's it. Because I tell myself, 'there are living things here.'" The future urban landscape would therefore be constructed on the basis of cohabitation. The landscape would be delimited according to species and population dynamics inscribed in ecosystems—in other words, a functional biodiversity. In this way, a functioning and beautiful animal landscape can emerge; an environmental aesthetic modified by science is adopted in the city by sensitizing the population to environmental questions. Certain individual species considered to be without value could be redefined as part of the natural urban environment.

In Paris, where new developments are progressing most notably in the context of "Biodiversity Plans," animal encounters in the downtown area continue to perplex. The question of enclosures is raised. The democratization of zoos at the beginning of the twentieth century allowed for crowds who were used to violence against animals to contemplate caged wild animals, even to harass them from behind the safety of the bars. Marking a separation is therefore essential to the interaction of humans and animals.[14] F (Paris center, environmentalist) says: "Animals still need an environment often specific to their species and origin. The city is still too hostile. Yes, you really do need enclosed walkways. You really do need enclosed spaces." How then can one envision a new sharing of territories? An ecologically oriented urban planning advocates for a dynamic opening-up of the city to its own environment.[15] The city is already an open space because of the networks linking it to surrounding territories, and because of the cycles and bio-physical-chemical processes (water, air, etc.) that must be considered by the urban planner, whose profession has been greatly redefined. Animals used to have no place in the city except as objects of display, curious zoological specimens, or pets. They are now, it seems, being reinscribed in urban spaces. What then is the place of animals in greenways?

## ANIMALS, A STORY TO THE ORDER OF *DIFFÉRANCE*?

The term *biodiversity* now covers all the roles that companion animals once played for humans. It is no longer about what is animal or vegetal, but about a collection of living species (including human beings to a certain degree) whose existence and preservation is justified most notably by the genetic resources at stake. Whether a particular species is wanted or not has little importance in terms of their mode of inclusion in social space. This expansion of new terminology to describe the living world has been particularly well received in urban spaces with the creation of greenways and their "blue" water counterpart. However, in promotional material about greenways, biodiversity is rarely mentioned in detail. The term is employed in a generic manner, conflating all the existing diverse species into one. This can generate two types of problematic perspectives. The first issue is related to how biodiversity is represented, as can be seen in figure 1.

In this image, biodiversity is represented by a kingfisher, a fox, a squirrel, and a butterfly: all animals that have positive connotations in the collective imaginary. Wolves, bears, or even the water-primrose (Latin name: *ludwigia*), a plant considered highly invasive in France, have no place in this statement promoting biodiversity.

The second issue is oversimplification. Indeed, when biodiversity is presented as a single entity, its management and conservation through ecological networks appear quite easy. But the reality is more complex, involving a much wider variety of animal and vegetal species, each characterized by a unique set of behaviors. The above image also puts humans, seen here in the figure of the young blond girl, at the center of the other species, all soft, gentle, and thoroughly anthropomorphized. Current discourse speaks quite differently of a global biodiversity that is endangered and threatened by human activities. In other words, the generic usage of the term *biodiversity* tends to erase differences: greenways are reduced to a spatial dimension, eliding the ordinary encounters between living beings in new urban spaces. On the one hand, biodiversity is misunderstood as beautiful, angelic; on the other hand, it is outlined by the spatial frame, transformed into an image, a totalizing description of an already attained happiness. This perspective contracts the space of greenways, reducing coexistences into a concentrated time.[16] And yet it is clear that with each animal encounter, the living world inscribes itself in

Informations dans les Espaces Régionaux des départements situés à :
Angers, La Roche-Sur-Yon, Laval, Le Mans, Nantes, Saint-Nazaire.

www.paysdelaloire.fr

*Figure 1.* Illustration from the front page of the activities program of "L'émoi des parcs" (For the Love of Parks), a month-long celebration of regional and natural parks in the Loire, September 2012. (Source: Conseil régional des Pays de la Loire, used with permission)

time. Furthermore, the notion of biodiversity that is committed to preserving endangered species is necessarily inscribed in a concept of unfolding time. Yet, the modes of implementing greenways, most specifically through the use of maps and chart diagrams, and also in the discussions themselves, tend to reduce greenways either to simple images without duration or to green infrastructures modeled after roadways or highways. This is a paradox. Sustainable development reintroduces duration and time as a developmental horizon that refutes planning policies oriented towards the simultaneity of a picture as a blueprint. The animal, and each "animot," is subsumed by the term "biodiversity," and yet, the particular animal, in each individual and encounter, refers back to the general animal who itself refers back to the city, the neighborhood, and domestic spaces, creating an infinite play of metaphorical, anecdotal, and associative significance that justifies the place of the animal in social and discursive space. The picture is what allows the developer to envision actions in synchronic time while the story creates duration. Only memory that retraces what has been read or heard and relates it to what will follow can create a sense of a coherent whole. The story and the cockroach pass in time, while the caged animal is more like a still-life painting with no possibility of escaping the frame.

## WHAT STATUS FOR ANIMALS?

If animals are granted a place in the city for the sake of biodiversity, how should we consider the modes of encountering "animots" in the context of such spaces? Do they not demonstrate the importance of new developments in animal ethics? The fight against the loss of biodiversity coincides with policies for nature reserves that dictate varying degrees of protection. The preservation of dynamic ecosystems outside these specific nature reserves requires a restructuring of planning principles: How is land to be divided? What would be the legitimacy, the scientific principles, and forms of knowledge supporting these divisions? Urban territories have their own set of issues: as the principal sources of biodiversity degradation, eating away at natural habitats, they are also "laboratories" (the term employed in urban ecology) for new interactions with the environment. Urban greenways are reminders of the sharing of spaces. Without going into the details of human and animal specificities, what might be the modes of sharing between living beings? Is it possible to think of a co-citizenship?

In France, there are three main types of democratic citizenship. First, there is the citizen, the product of deterritorialized and abstract relations to the public space, a condition of geographic equality, and a representative of elective democracy. Second, there is the producer, the result of the shift in forces of production from nature to humans; and third, the inhabitant, the "backward-looking" representative of a locality often opposed to the "modern" forces of the nation. To some extent, the city is the symbol of modernity, while the countryside, even if it is the backdrop to this urban modernity in progress, remains the refuge of anti-modernity, anti-cosmopolitanism activists (Larzac, hunters, NIMBY, etc.) As a representative of participatory democracy and a figure of the local, the inhabitant is associated with the rural. While these are three distinct figures, it is possible to consider all living beings as inhabitants by invoking corporeal attachment to territory. As guarantors of environmental policies that aim for a quality of life via the quality of land, inhabitants represent a mode of existence in the world, a presence on the earth.

And where does the animal fit in? This is not a question of revisiting the entirety of animal ethics. Environmental ethics do not change in any essential way the nature/culture relationship and do not make any less specific the human domain.[17] Even if holistic ethics appear to consider nature as a complex and dynamic whole that preserves the interactions between living beings, they only extend the fields of expertise for the administrator and the informed scientist. Animal ethics tend to reconsider the place of animals as rights-bearing subjects and sensory beings; yet these ethics also allow little place for species whose members are not easily differentiated as individuals (some even reproduce like plants). It is only the relationships with animals that most resemble human beings that are taken into consideration. Furthermore, this form of ethics does not take into consideration the context of social practices in which animals are found. How can we speak of the animal and make it a rightful subject modeled after humans, when work is being done to deconstruct human supremacy and to erase human universality for the sake of increasingly singular individual identities? The separation between environmental and animal ethics reflects the deep divide between holistic environmentalists and animal-rights defenders, who tend to be rather individualistic. More recently, the "care" ethic associated with environmental ethics emphasizes the idea that environmental degradation concerns all of us.[18] We have little choice but to adopt some kind of "care" ethic, since environmental dangers affect our living conditions. The image of an environment that has degraded, perhaps irreversibly, transforms ethical relations with nature. Among these relations, there is firstly the relation to our immediate surrounding environment that requires us to respect it for the sake of the quality and setting of our life; then there is the ethical relation to a distant nature, sometimes brought close through the role of media. What is also born of this spectacle is a feeling of shame or guilt for not having taken care of nature. These feelings are equally associated with future generations and the need to take into consideration the fragility of existing populations and environments. This fragility makes our obligation all the more evident. We are all linked because the earth is one ecosystem; fragile beings are all linked and are sometimes necessary to the survival of those that are less fragile. Earth, as an ecosystem or representation of an ensemble of interconnected natural workings, shows us a chain of living beings and natural elements, all difficult to put into a working model. These workings naturally concern human beings, but also all other living beings and natural entities defined as such by their historical relationship with human beings.

The choice does not extend from a contract since there is strictly speaking no moral obligation; planetary cohabitation with all sorts of entities supposes a redefinition of the conditions involved in community living. The care ethic (concerning the care given notably to precarious individuals) applied to environmental ethics allows us to conceive of the care afforded to fragile living beings needing support. In this sense the care ethic facilitates the articulation of an ethics of group solidarity, either close or far away.[19] Progressing in stages, one can first say that the animal participants of biodiversity present a kind of obligation. In this way we must protect the living world as well as ourselves. It is all the more true in the case of greenways; this ecological development of new sites for communal living is representative of a new obligation towards solidarity. This obligation emphasizes relationships with constructed and natural environments that demonstrate successful cohabitation and co-adaptation.[20] Human and non-human beings as representatives of biodiversity are expected to find the appropriate distances from each other, while also putting in place specific modalities of interaction.

Or perhaps we should speak of encounters rather than interactions. What is implied, in effect, is managing encounters within designated spaces, but without fences to separate human and animal bodies. Considering the example of pigeons and insects in the urban environment, the encounters have symbolic, imagined, and concrete implications. The encounter entails finding oneself face-to-face with, and equal to, an animal while also contemplating the singularity of that particular animal. One day, while walking in the sewers, I found myself in front of a band of rats. The headlamp had illuminated the first one and, while looking at me, he stood up on his two back legs; the phrase "Doctor Livingstone, I presume?" immediately came to mind. I was a stranger standing before the indigenous race of the sewers. The encounter involves, therefore, accepting that verbal exchange is not the only way of experiencing such meetings. It is an aesthetic exchange that appeals to sensations, feelings, and the beauty of the living being, and in which the interagency between a human being, a rat, and a sewer is acted out like a picture described from all points of view.

It is likely, however—and my work on greenways serves as evidence—that this obligation will only be justified if numerous criteria are taken into consideration. Species believed to harm the health of human beings, to be invasive, or repulsive in appearance are simply not tolerated. Such a distinction brings into play aesthetic criteria and values associated with cultures of nature. The real-life ethic therefore illustrates a relation of solidarity with natural and constructed environments that creates its own acknowledgment of itself and of the other living beings by and through the environment. Human beings cohabit and co-construct with the elements of their surrounding environment that they define as such, and that define them in return. This process happens most notably through aesthetics, landscapes, and narratives.

## CONCLUSION

The city dweller experiences nature in a changing system, according to the logic of the city, but equally according to aesthetic and ethical criteria expressing the diverse sensitivities of living beings. Broadly, the idea of the greenway refers to and reconnects the city dweller with nature. The greenway highlights the spatial and temporal linkage between human beings and other living beings. However, animals are privileged in this environmental construction (in comparison to vegetation) and play an intermediary role in its constructed elaboration. In a more general sense, animals can help structure human relations to nature within the city. They allow us to tell the stories of their existence within the environment, and to give these stories form and meaning. Is this not the key to a new nature? Among the possible new associations between nature and policies, one of them could award to animals the role of actual "citizens": animals exemplify the achievement of a successful biodiversity. These territorial associations and layouts would then award a place of particular importance to the idea of habitation and therefore cohabitation. To reenvision the place of nature, of different natures in our various cultures cannot simply mean adding a new actor or group of actors to our democratic process.[21] It entails, rather, a more profound modification of how our democracy is enacted. The citizen has become an inhabitant. From this perspective, in full awareness of the difficulties of local democracy, how can we postulate the importance of territorial attachments?[22] The evolution of links between nature and artifice no longer challenge the hybrid quality of species and spaces; a new form of environmental art is what is now required. In all truth, the policies of sustainable development emphasize that the city is not (or no longer) an enclosed space, but a mode of living that is based on the use of natural resources.

## NOTES

1. Decreed on July 12, the Grenelle 2 law proposed "an environmental engagement on the national level (ENE)" that coincided with the application of some of the commitments of the first *Grenelle Environnement.* The idea of greenways was notably foregrounded. (Translator's note: the French term for greenways is "trames vertes.")
2. The term *biodiversity* as a contraction of biological diversity refers to all the living constituents of the biosphere. More than a simple inventory of the present species, biodiversity strives for a complete study of all that is living (bacteria, fungi, vegetation, and animals) organized into three different levels: genes, species, and ecosystems. In this way, biodiversity includes diversity between and within species, as well as diversity of ecosystems.
3. See Giorgio Agamben, *L'ouvert: De l'homme et de l'animal* (Paris: Payot Rivages, 2002) [*The Open: Man and Animal,* trans. Kevin Attell (Stanford, CA: Stanford University Press, 2003)].
4. Greenways differ from traditional greenbelts insomuch as they are conceived as a green ecological network, interconnected rather than separate strips or dots of parks or natural spaces.
5. Nathalie Blanc, *La nature dans la cité* (doctoral thesis, Université de Paris 1 Sorbonne, 1996).
6. As proposed, for example, by Maurice Godelier, *L'idéel et le matériel: Pensée, économies, société* (Paris: Flammarion, 2010).
7. Translator's note: "chat libre" in French. The English "feral cats" does not quite convey the same sense.
8. Jacques Derrida, *L'animal que donc je suis* (Paris: Galilée, 2006), 65 [*The Animal That Therefore I Am,* trans. Marie-Louise Mallet (New York: Fordham University Press, 2008)].
9. Nathalie Blanc, *L'animal dans la ville* (Paris: Odile Jacob, 2000).
10. *Evaluation des trames vertes urbaines et élaboration de référentiels: Une infrastructure entre esthétique et écologie pour une nouvelle urbanité.*
11. See Blanc, *L'animal.*
12. Ferdinand Hayden uses William Henry Jackson's landscape photos and Thomas Moran's printed engravings as supporting evidence in favor of preserving Yellowstone National Park, created in 1872.
13. See Bernard Kalaora, "Naissance et développement d'un loisir urbain: La forêt de Fontainebleau," *Études rurales* 83 (1981): 97–115.
14. See Éric Baratay and Élisabeth Hardouin-Fugier, *Zoos: Histoire des jardins zoologiques en occident (XVIe-XXe siècle)* (Paris: La Découverte, 1998), 201.
15. See Jennifer Wolch, "Zoopolis," in *Animal Geographies: Place, Politics, and Identity in the Nature-Culture Borderlands,* ed. Jennifer Wolch and Jody Emel (New York: Verso, 1998), 119–138.
16. Jacques Derrida, *L'écriture et la différence* (Paris: Seuil, 1967), 42.
17. This has been argued notably by Pierre Descola, *Par-delà nature et culture* (Paris: Gallimard, 2005).
18. See Catherine Larrère, "Care et environnement: La montagne ou le jardin?," in Sandra Laugier, *Tous vulnérables? Le care, les animaux et l'environnement* (Paris: Payot, 2011), 233–263.
19. It should be noted that care of a living being, animal or vegetal, that is within proximity of the caregiver is often classified as a feminine activity. See Blanc, *L'animal,* and "La place de l'animal dans les politiques urbaines," *Communications* 74 (2003): 159–174.
20. Elinor Ostrom, *Gouvernance des biens communs: Pour une approche nouvelle des ressources naturelles* (Bruxelles: De Boeck, 2011).

21. On ecological democracy, see Dominique Bourg and Kerry Whiteside, *Vers une démocratie écologique: Le citoyen, le savant et le politique* (Paris: Seuil, 2010).
22. On democratic participation, see Loic Blondiaux and Jean-Michel Fourniaux, "Un bilan des recherches sur la participation du public en démocratie: Beaucoup de bruit pour rien?" *Participations* 1 (2011): 8–35.

# Wild, Domestic, or Technical
## *What Status for Animals?*

MARIE-HÉLÈNE PARIZEAU

*Translated by* STEPHANIE POSTHUMUS

In philosophy, the animal has often served as a foil for the human. Human specificity has largely been constructed in contrast to the animal, most significantly since the rise of Western modernity, which radically separated the human from nature and made the animal an object. Several contemporary philosophers are now interested in the animal in and of itself, in its capacity to suffer, and in its relationship to a world that it is capable of creating. The issue of the status of the animal has since given rise to a number of varying responses. The present article will analyze various branches, first those that have appeared in Anglo-Saxon circles, more particularly North America, and second those that have been developing in Francophone Europe, more particularly in France and Belgium. This comparative angle aims to show that the philosophical consideration of the animal question is related to the dominant conception of nature, culturally constructed and historically situated. Despite their differences, these new modes of thinking about the status of the animal all constitute ways out of a modernity experienced as destructive and alienating as much for the animal as for the human being.

### THE ANIMAL QUESTION FROM A PHILOSOPHICAL PERSPECTIVE: DEFINING THE HUMAN AND ANIMAL CRUELTY

Since Greek Antiquity, philosophy has been interested in the animal as a way of defining, either in continuity with or in opposition to, the essence of the human being or human specificity. Interest in the animal was indirect: the animal was not analyzed or understood for itself, in its general nature, or in its relational specificity. For example, if Aristotle defined the animal as having a sensitive soul, it was in order to better elevate man as a "rational animal" in the hierarchy of living beings. In this way, man was part of nature, in continuity with it, yet still the most perfect of all living beings. He displayed his humanity in his own environment, the city, by becoming a citizen.

As for Descartes' well-known machine animals, they were considered radically different from man in that they did not possess speech or rationality. Incapable of thought, these animals moved about like machines in a mechanical universe where the laws of nature were written in mathematical language.[1] In this way, Descartes inaugurated modernity's rupture by separating the soul connected to God from the material body, and the human being from the world that surrounded him. Reason became the main method by which the subject could know the object.

However, scientific knowledge can sometimes use reason to play its own tricks, as when Darwin brings humans back into nature in the nineteenth century. This science of natural evolutionary theory posits man as a product of nature, and not a divine creation made in the image of God. Drawing inspiration from the work of embryologists, Darwin shows that "man is descended from apes"[2] and is not the product of an animal series that gradually culminates in man as the state of perfect being. On the contrary, man is simply the most recent being in this evolutionary history; he is in no way specialized as the final product, and he could very well be surpassed in the future. For Darwin, hominization happens through many different stages of acquiring specific traits, some of which are already present in the great apes, such as emotions and laughter. Humans and animals have the same sensory organs and thus will feel the same sensations. There is no fundamental difference between humans and the highest functioning mammals, even at the level of intelligence. It's simply a question of degree. Darwin was thus very attentive to finding the human in the animal.[3]

The ethical question of animal treatment is related to the way in which the dominant philosophy of a given time imagines the respective places of the human and the animal. Animal cruelty has certainly been a concern of animal ethics throughout the history of philosophy. This moral preoccupation can be found from Pythagoras to Porphyry, and is taken up in varying degrees by the main religions.[4] The reasons for this ethical concern for animals can be very different. In the sixth century BCE, Pythagoras believed in metempsychosis, according to which the soul changes bodies after death and thus can find itself in the body of an animal. This possibility necessitates that animals be given a particular ethical status as well. Pythagoras recommended frugality and vegetarianism. Killing an animal, he claimed, constituted an incitement to kill a man. As for Plutarch (46–125 AD), he put forward similar ethical principles using a different philosophical foundation, that of reason being shared by humans and animals. In the second century, the neo-Platonist Porphyry extended the principle of justice to animals in the name of a shared *logos* between humans and animals. Respecting others and practicing justice, his ethical attitude was one of abstinence, opposed to the search for pleasure.

The moral question of animal cruelty is also discussed by philosophers of modernity. At the height of rational conceptions of the human being, Kant explains that animals are things that can be appropriated, in opposition to humans, who are autonomous, rational subjects with their own dignity. Humans must, however, take specific precautions with animals. Although humans can use animals, even kill them, they cannot make them suffer. Violent treatment of animals dulls the natural predisposition to feel compassion when faced with human suffering, according to Kant.

In the scientific laboratories of the nineteenth century, there was not much room for animal compassion as animals became the object of experimentation and vivisection. In his *Introduction à la médecine expérimentale* (1865), Claude Bernard posits the right of knowledge over the rights of the animal. In the medical sciences, it is more ethical, asserts Bernard, to first experiment on animals rather than on human beings. The first societies for the protection of animals, which began appearing in 1824, protested this treatment of animals (vivisection, animal fighting, etc.), and their actions led to the creation of laws first in England, then in France.

It is in this general context that the question of animal suffering is taken up by the moral theory of utilitarianism in the nineteenth century, by Jeremy Bentham and then by John Stuart Mill.[5] Because animals are sentient beings, utilitarianism places them within the moral community, the criteria for morality being the capacity to feel pain and pleasure. The good (the useful) is determined by calculating the results of applying this criteria—in other words, by maximizing pleasure for the greatest number. But it is not until the birth of environmental ethics in the twentieth century, and more specifically in the 1970s, that this moral tradition, transformed by the linguistic turn, reemerges under the name "animal liberation," the title of Peter Singer's book published in 1975. This marks the start of animal ethics, with debates first taking place in North America and emerging at the same time as environmental ethics. While the latter critique Western modernity, or at least some of its characteristics, animal ethics remain rooted in the debates that characterize modernity's classical moral philosophical tradition.

## CONFLICT BETWEEN ANIMAL ETHICS AND ENVIRONMENTAL ETHICS IN THE UNITED STATES

The originality of Peter Singer's approach lies, on the one hand, in the fact that it takes up a renewed utilitarianism, and on the other hand, in the fact that it develops a practical dimension denouncing concrete and real ethical problems. On the theoretical level, Singer asserts the equality of humans and animals capable of feeling pain and pleasure; as such, the utilitarian's calculation of the individual must take into account the interests of all those who might be affected by the consequences of that individual's actions. On this basis, Singer denounces discrimination against animals and the immoral treatment inflicted on them, such as animal experimentation, factory farming, bullfighting, because this treatment generates suffering and does not take into account animals' interests.

The well-known theoretical quarrels in moral philosophy between deontological ethics and consequentialist ethics[6] resume with even more fervor around a new issue: the moral status of animals. In response to Singer's utilitarianism, Tom Regan publishes *The Case for Animal Rights* in 1983. Regan defends moral rights for animals, which would require the creation of stricter legal rights. His argument is based on the fact that animals are all living beings, they all have inherent value, and they all possess value equally. They are not defined as receptacles of experience (sentient beings, according to Singer). Given the animal's inherent value, a principle of respect is necessarily established from which a principle of justice towards animals can be derived. The animal's right to fair treatment—Regan speaks of respectful treatment—is morally admissible and attributable. Animals thus have basic rights—in particular, the right to be treated with respect. Unlike utilitarianism, this reasoning does not appeal to the idea of general well-being that is at the heart of the principle of utility. Regan follows neo-Kantianism in terms of a legal and ethical foundation, but goes beyond it by proposing a deontological ethics that includes animals in the name of their inherent value. He raises an important ethical and legal question about animal rights that becomes the subject of a prolific body of literature in the Anglo-Saxon world.[7]

As can be seen in a large number of follow-up publications, the debate between Singer and Regan becomes the required reference for any further reflection on the question of animal being.[8] Revisited again and again, their theoretical quarrels have given rise to ethical questions that are more and more specialized, or, on the contrary, that are situated within a larger philosophical context. Once established, this debate gives the impression—and this is what makes it interesting on the philosophical level—that the question of animal ethics remains on the margins of environmental ethics, separate from the moral questions that preoccupied North American environmental ethics that were on the rise in the 1970s. In a vindictive article published in 1980, J. Baird Callicott accentuates the divisions between environmental ethics and animal ethics by positing them as radically separate.[9] While environmental ethics are holistic, he argues, animal ethics are atomized, focusing either on the animal's capacity for suffering or on animal rights. Only interested in the ways in which humans treat domestic animals, these animal ethics make a case for animal rights and animal liberation using a logic similar to that used to give rights to women or freedom to black slaves. This extension of rights is not sufficient for establishing an environmental ethic, according to Baird Callicott.

Drawing on Aldo Leopold's "land ethics,"[10] Baird Callicott argues that the principal ethical issue is the preservation of the beauty, stability, and integrity of the biotic community in its entirety, including ecological equilibrium (predation, food chains, etc.). Wild animals and plants in this natural environment possess a special place in nature: they are the fruit of natural evolution. This status cannot be attributed to domestic animals, for they are the product of human technology. Domestic animals are human creations, living artifacts, extensions of human activity. The ethical argument of a domestic animal's "natural behavior" is thus devoid of meaning, because these animals are unable to return to a wild state due to the artificial selection of which they have been the object. Moreover, adds Callicott, the flocks of sheep or herds of cows do at least as much damage to natural ecosystems as off-road vehicles.

The demarcation line between animal ethics and environmental ethics that can be found on the level of the animal's moral status is rooted in a deeper dichotomy between nature and artifice: on the one hand, the domestic animal, who is a product of human culture and technology; and on the other, the wild animal, who has inherent value as the natural product of evolution. The true, authentic animal is the wild animal who perfectly reflects the image of its environment, the wilderness, a nature without humans. This nature, conceived of as "wild" or "pristine" before the birth of hominoids, has been imagined from two different angles. The first, scientific, represents nature as a product of Darwinian evolution, while the second, religious, posits a perfect nature created by God as an earthly paradise. Although described as a garden, this paradise does not become one until Adam and Eve make their alliance with God. Before creating humans, God created an ordered nature following the order of the species—a nature that was free of the agrarian work that God later commanded humans to do.[11] Wilderness thus reveals itself as a culturally inflected concept with many scientific and religious connotations.

The American valorization of wilderness that is part of biocentric environmental ethics represents a sociohistorical and a philosophical phenomenon. It can also be linked to the history of the beginning of environmental ethics as constituted by American philosophers since the opening act of the dual, contradictory conception of nature found in Pinchot (preservation and use of nature) vs. Muir (conservation of nature), an opening act that finally found its resolution in Leopold's work on biotic communities.[12] Roderick Nash's 1967 book profoundly influenced the sociohistorical understanding of the construction of wilderness in the United States.[13] As Nash explains, the first European settlers arrived with a Christian concept of nature. Man was the guardian of nature, but also exercised a form of domination over it. The idea of "wild" had a negative connotation, meaning a dangerous and evil place. The gradual colonization of American land all the way to the Pacific came to a close in 1890 with the end of the frontier. There were no more wild places to conquer, and industrial development was radically modifying previous ways of living and former environments. The first national parks were established between 1872 and 1916 (this latter being the date of the creation of the National Parks Service) as a means to preserve the exceptional landscapes and what was left of wildlife. Nash explains this movement by pointing out who exactly accorded such value to wilderness: an East Coast urban elite, living in the city and undergoing the effects of industrialization, and so believing that these wild places might heal the harms of modernization.

This romantic movement was incarnated in the emblematic figures of Emerson and Thoreau, who praised and justified a natural way of life and searched for a harmonious relationship with nature. But as Max Oelschlaeger points out, this Romanticism had roots going back to Europe with Rousseau, who rejected culture's corrupting influences on natural man.[14] One might also cite the importance of German Romanticism at the beginning of the nineteenth century, which embraced *Naturphilosophie* as the foundation of its worldview, and which refused to isolate man from the rest of the cosmos: the human being was in the world; he inhabited it.[15]

The concept of American wilderness thus echoes other branches of thinking about nature within Western, and more precisely European, philosophy. But from a historical perspective, it is clear that this concept becomes the main value for policies of environmental conservation—in particular, parks—first at the U.S. national level and then at the international level. This value is affirmed in the 1964 Wilderness Act,[16] reinforcing the representation of pristine nature, untouched by humans, and crowning at the same time the wild animal as the incarnation of the category "animal."

To summarize, American debates about the status of animals follow two distinct theoretical lines. The one expresses an ethical consideration of animals, for the most part domestic, in line with important modern moral traditions—Kantianism and utilitarianism. The other, from an ethical position critical of certain aspects of modernity, proposes a holism in which the wild animal (wolf, polar bear) becomes a symbol and a representation of an intact nature that must be conserved at all costs. Only a few voices have indicated a way out of this dichotomy. Mary Migley is a notable example. She argues that humans and animals have been living together in mixed communities for thousands of years. This social coevolution has resulted in the sharing of several common emotions: sympathy, trust, compassion, love. Animals and humans are in a dynamic, evolving interspecies relationship constructed by society. This perspective resonates more with that of European thinkers, and more specifically with that of Francophone philosophers.

*Figure 1.* Yellowstone National Park, ca. 1895–ca. 1915. (Source: Robert N. Dennis collection of stereoscopic views; Public domain—New York Public Library)

## FROM THE RELATIONAL ANIMAL TO A CARE FOR NATURE: FRENCH AND BELGIAN PERSPECTIVES

While the animal question in the North American context has a history with two identifiable main branches—animal ethics and environmental ethics—the European context, especially Francophone, does not lend itself so easily to this type of analysis.

Around the 1990s in France, the animal question was once again taken up alongside movements for animal protection that had been active since the nineteenth century. These movements, in particular the International League for Animal Rights, succeeded in having the Universal Declaration of Animal Rights solemnly declared at UNESCO's headquarters in Paris in 1978. This text's ethical aim was to affirm the right of all animal species to exist, as well as to demand that humans respect these lives. It denounced, in particular, acts of cruelty towards animals, while affirming the right to a death free of suffering when death was necessary, asserting the need to limit animal experimentation through the use of replacement technologies, and calling for the banning of animal exhibitions and shows. Not only concerned with domestic animals, this text highlighted the link between polluted or destroyed environments and "biocide" or crimes against life. More fundamentally, the idea that "all animals have the right to be respected" was clearly supported. This declaration was highly criticized: some saw it as a misuse of the Declaration of the Rights of Man (1789) and responded that only humans were free from animal determinism;[17] others considered, similarly to Claude Bernard, that animal experimentation was essential to the advancement of knowledge, and that limiting it was synonymous to abandoning progress.

The polemic then migrated to the terrain of jurists who began examining the nature of animal rights and the ways in which an animal can be a "subject of the law." This question was linked to another legal issue, that of the patentability of living beings, or more specifically, of transgenic animals in the 1980s and then of cloned animals in the 1990s. The grip of patent rights began to grow progressively tighter around the concept of "biological material" that included, somewhat vaguely, plants, animals, and humans. Jurists therefore reintroduced these three categories, distinguishing them in order to authorize, or not, certain uses. In doing so, they affirmed the non-patentability of human beings "as such," separating once again humans from all other living beings and affirming humans' unique dignity. But given that animals are sometimes similar to humans, jurists constructed different solutions in order to take into account animal specificity. The evolution of jurisprudence in Europe as in the United States nevertheless continues to keep animals in the category of legal things;[18] they can still be the vehicle of a patented invention. However, transgenic animals cannot be the object of a patent if the genetic modification induces animal suffering, unless the genetic modification constitutes a substantial benefit to human and animal medicine.

While European law seems to have made its ruling and refused rights or a separate status to animals, animal suffering still constitutes an ethical limit that is recognized by the law and that is applicable to research laboratories. Georges Chapouthier, philosopher and biologist, as well as Jean-Claude Nouët, doctor and president of the French League for the Rights of Animals, have been the most passionate defenders of this position, which is supported by the claims of a number of associations for the protection of animals in France, Belgium, and elsewhere. These associations continue to regularly and vigilantly denounce practices that are cruel and disrespectful to animals (factory farming, hunting, zoos, corridas, animal abandonment, etc.).[19]

From a philosophical perspective, this animal-rights movement illustrates that animals are not categorized as wild, domestic, or artificially made. There are no separate categories. The animal is a living being, regardless of the uses that humans make of it or the relations that humans have with it. Its unique status translates into a demand for dignity and rights in terms of how it is treated: at the very least, without cruelty. Chapouthier, in order to arrive at this position on animal rights, defends a continuist position according to which the human is also an animal, and yet not just an animal.[20] He draws on biological and ethological research that illustrates the existence of a proto-culture (primates' sense of beauty), the use of a proto-language that includes abstract thought, and the presence of a consciousness that expresses personal lived experience and a proto-morality. For the human being, explains Chapouthier, it is from a sense of aesthetics that a sense of morality grows, allowing humans to treat animals and the environment with solidarity.

In Belgium, the foundations and organizations for the protection of animals have also contributed to changing certain ways of treating animals, in particular laboratory animals.[21] As for new developments in philosophy, much work has been done on the changing relationships between humans and animals in our current biotechnological era.[22] How do we imagine a new alliance with animals when these latter (e.g., transgenic mice) are used and modified for scientific research ends that can be economic (patents), cognitive, or pharmaceutical? A tension arises in scientific research between an extreme instrumentalization of animals (transgenic animals transformed so as to manifest certain diseases), on the one hand, and on the other hand, the ethical responses that attempt to minimize the suffering and discomfort of laboratory animals by using replacements (in vitro cultures, virtual modeling), by reducing the total number of animals used, and by refining pain relief (analgesics).[23] In their current practices, scientists no longer consider an animal to be a mechanical instrument, a laboratory reagent, or a bioreactor, but rather as a living being whose specific sensibility they need to understand.[24] The concept of a "model animal" has changed greatly, but laboratory animals still remain at the service of human ends. Despite attempts at a new alliance, the animal's status remains constructed through weak anthropocentrism.

This openness to, and search for, a new alliance or relationship between humans and animals that goes beyond the categories of modernity are characteristic of French and Belgian thinking that is now emerging in the work of philosophers, scientists, and animal-rights advocates. We will rapidly summarize the thinking of four French and Belgian philosophers—Élisabeth de Fontenay, Vinciane Despret, Dominique Lestel, and Florence Burgat—who, from slightly different perspectives, explore the opening up of the human/animal relationship, either by calling into question various practices or by searching for a new ethical and philosophical foundation to change our perception of animality.

Élisabeth de Fontenay supports the animal cause through her critique of the concept of *le propre de l'homme* ("what is proper to the human"), a concept that has defined the metaphysical tradition's opposition of the human and the animal. De Fontenay examines the borders of anthropology by deconstructing, similarly to Derrida,[25] modern humanisms that glorify human specificity, whether this be in the form of language, reason, thought, awareness of death, or their corollaries: the Subject-Individual, the Culture-Nature dichotomy, the History-Progress or Science-Technology combination. Modernity either condemns animals to be objects transformed through industrial production and science, or endangers them as a species because of human activities and influences. The technologies of animal production in particular lead to the denaturation of the animal's animality. De Fontenay speaks of human barbarity to describe the treatment of factory farm animals.[26] When she critiques the massive slaughter of bovines during the mad cow disease epidemic, she does not hesitate to speak of "extermination."[27] Animal treatment has become a symptom of productivist, technical, mercantile, humanist Western civilization that rejects the baseness of our ways of life. De Fontenay appeals to a sense of pity, or rather to a sense of goodwill, a benevolence towards animals as a means to creating a community of living beings in which, as Derrida would say, "the animal watches us/regards us/concerns us."[28]

This animal gaze/regard/concern has become a source for new ways of thinking about the human/animal relationship. French philosopher and phenomenologist Maurice Merleau-Ponty's openness to animality has inspired many ethnologists, psychologists, and philosophers to adopt a new position regarding human/animal relationships.[29] Belgian philosopher and etho-psychologist Vinciane Despret wants to open up our understanding of the animal by asking questions that start with the animal in its relationship to humans. We must change our perspective, she explains, as well as the way we conceive of the animal and the way we do science. In addition, we must revisit our prejudices concerning farm animals. Following the line of thinking of Jakob von Uexküll, Despret proposes seeing the animal from its own sensorial and relational modes that constitute its world (its *Umwelt*). This animal *Umwelt*, as Canguilhem explains,[30] consists of a milieu of behaviors that are specific to the animal. To change perspectives, it is necessary to enter into a relationship with the animal in its own environment, to make use of its sensibility, which is understood as a form of availability to the world or to certain events that happen there. This availability to the animal's world can be seen when researchers bring their responsibility and empathy to bear on working with animals, and thus change the way of doing science.[31] By examining the suffering inflicted on animals and by better understanding animals' adaptive modes and well-being, researchers and farmers can change their understanding of the relationship they have with animals, not in order to think for the animal, but to think with the animal, and thus correlatively change certain practices and create a world together.

French philosopher and ethologist Dominique Lestel explores ethological studies of animal behavior that call into question the divisions between the human being and the animal. Scientists continue to debate the continuity of animal and human culture—primates do use technical tools—as well as the distinction between the animal *Umwelt* and the human *Umwelt*. The question of the animal's "interior life" remains open. In his reflections on animal culture, Lestel examines how culture emerges from the phenomenon of life[32]—Lestel's reference to Hans Jonas is explicit—in a process of individuation from the collective. The complex expressivity of individualization goes beyond the adaptive functionality of the species. From a continuist perspective, Lestel explains that the human being, like the animal, evolves at the boundary between nature and culture; the animal is thus subject and singular. Animals and humans form "mixed communities" in which relationships with certain species are privileged over others.[33] This living together in hybrid human/animal communities constitutes a form of association where interests, affects, and meanings are shared, argues Lestel. Mixed communities that have their own particular partnerships and that have always existed in a diversity of cultures now present more of a problem in our modern world. Widely mistreating animals, or on the contrary according them the status of subject, also has consequences for the modern world. Rethinking relationships with domestic animals requires that we rid ourselves of the old ontological habit of human exceptionalism and widen our thinking anthropologically to reconsider the notion of person.

As for Florence Burgat, her philosophical work aims to shift the focus from animal suffering (as in Singer's utilitarianism) to a better understanding of the animal mind, or rather the capacity of animals to think. Continuist positions, she explains, see the animal on a continuum with the human species, but as still having something less than the human. How can we understand this problem differently and understand the different forms of animal life? Drawing on phenomenology's attempts to understand life in its biologically rooted context, Burgat examines the concept of animal existence. From this perspective, the animal is an *existant*, that is, the animal experiences its own existence, as the subject of a story (and not just a biological path), with its own life of relationships, with its own coordinated yet indeterminate behaviors, and with its own emotions (boredom, anxiety). Burgat explains that conscious life is determined by the style of the species and manifests itself as a presence of self, since behavior is not determined in advance.

From Husserl to Merleau-Ponty to Canguilhem, other philosophers have prepared the conceptual ground for asking similar questions. In this way, subjectivity is no longer defined in terms of a reflexive layer (the reflexive consciousness that represents experience or later remembers it), but rather includes the pre-reflexive layer of life. Burgat distances herself,

***Figure 2.*** Engraving by A. F. Bauduin of Versailles seen from the Orangerie, "Veue du chasteau de Versailles du costé de l'Orangerie," 17th century. (Source: Bibliothèque nationale de France)

however, from cognitive studies on the representational mind of animals.[34] A phenomenological approach means leaving the animal to be free in its environment, since the animal possesses its own world. Animals are part of our world; we encounter them. We need to do the work of reconstruction using analogies drawn from the body and the senses—even if the world of rodents is less familiar than the world of primates, with whom human beings share more corporal analogies.

This analogical work leads to ethics because the knowledge and recognition, even if only partial, of animal worlds reveal the depth of their existence. Human beings cannot continue to use animals the way they presently do and simply satisfy themselves with reducing animal suffering. Burgat concludes that there is no replacement of one animal life for another: animals only have their one life to live, each life is unique, each one has a biography. Animals have an attachment to their life, a desire to live that we share with them. From this radical position, Burgat requires that we rethink in its entirety our moral behavior towards animals.

So why do these philosophers insist so strongly on the redefinition of our relationship to animals and the rewriting of the terms of the domestic contract? French philosopher Catherine Larrère explains that the present issue relates to becoming familiar with animals, to once again living in society with them. This human/animal socialization has always been made up of exchanges (affects, services), reciprocal obligations, but also unequal relationships. The issue at hand in today's mixed communities is one of communication; in the modern context, many domestic animals (in particular factory farm and laboratory animals) are treated as machines. This treatment is far from the moral imperative "to care for" that has been a constant in the relationship that Europeans have maintained more generally with nature.[35] In Europe, nature is not imagined as wilderness, because wild species have been replaced by a nature that has been landscaped and managed for many centuries now.[36]

The human being, if part of nature, has nevertheless made "good use" of it. To "make good use" is part of a relational mode of living with the environment; animals enter into this woven fabric of relationships. However, the standardization of systems of production and technology, the valorization of the "American way of life," the globalization of commercial exchanges, have greatly modified this relationship, which previously allowed for a relatively balanced interaction with a landscaped and inhabited nature, and which also led to a familiarity with animals. Today, these philosophical reflections, and more specifically phenomenological reflections, about the animal represent possible pathways for rethinking all of our modern human relationships, not only with animals but also more generally with the environment, the lived milieu.[37]

## THE ANIMAL AND TECHNOLOGY: THE STAKES OF HYBRIDITY

Only a few of these philosophers have considered the question of modern technology and its capacity to modify animal and even human species. The underlying philosophical issue is that of the hybridity of the living being, in this case the animal, and the role of the technological transformation. Does technology generate an animal artifact and thus bring the animal completely into the sphere of human culture? Or is technology, as Aristotle thought, simply an imitation of nature, so that the modified animals are part of the potentialities of nature in the same way as some of nature's monstrosities? In short, the nature/culture dichotomy raises its head once again. If the animal modified by technology (techniques of selection and enhancement, transgenesis, cloning or chimeras, etc.) is considered a cultural artifact, it is generally done so in opposition to the wild animal. From an anthropocentric perspective, ethical considerations would follow utilitarian rules (animal suffering, general conditions of well-being, cost-benefit analysis, etc.).

In public and private laboratories—whether these be in Europe, North America, or Asia—the techniques used on animals are already producing cloned cats and sheep, transgenic patented mice, transgenic pigs destined to be used for xenotransplants in humans, mice chimera whose brains are made up of a majority of neural human cells, etc. A catalog of monstrosities for some, a set of tools of knowledge for others, a hope for financial profit for many, these "animals" raise new ethical and philosophical questions as to their status and their specificity.

The philosophical question of the boundaries between the human and the animal is once again becoming a complex issue. In terms of the technical, one might ask, for example: if one can transfer a human gene into an animal, why not the inverse if this is equally beneficial and useful? Western societies accept the idea of grafting an organ like a pig's liver into a human because so much research is proposed and financed in this area. However, is it acceptable to graft cells or tissues or stem cells from human embryos into a primate embryo at the risk of "humanizing" it? Must there be an ethical limit on these "mixtures"? Moreover, what does it mean for an animal to be "humanized," or inversely for a human being to be "animalized"? Ethically speaking, is it necessary to maintain, and thus define, either the "essence" of the animal or "human nature," as Jürgen Habermas argues?[38] Or, on the other hand, should we be receptive to all these transformations of life that technology has made possible and begin establishing new relationships in the face of this generated hybridity?

Many scientists defend this latter point of view, citing the plasticity of nature: the laws of nature themselves allow for modifications, as well as the exchange of genes, tissues, and organs. Moreover, they explain that even the concepts of species and genes are shifting and represent more of a convenient vocabulary to describe an almost infinite plurality of forms and functions. The artificial thus becomes "natural" and scientifically justifiable.[39] The living world serves as a model for the technological so that we simply have to better define the ethical relations we would like to have with these modified animals, to redefine our alliances with them. Such a position is entirely compatible with a phenomenological position that would be attentive to the modified animal, and that would care for it in a mixed community.

Those who have embraced artificiality have often adopted the idea of human/machine hybridity (the cyborg)[40] as a sort of scientific ideology during the development of nanotechnologies in the early twenty-first century.[41] In the wake of the cybernetic movement of the 1950s, some of these thinkers also argued for a "post-humanity" in which the necessary artificial conditions would be created so that virtual "life forms" could give rise to artificial intelligence as a new property. Made possible by computing and computer science, the living being/machine interaction would open the door to a radically new diversity that would obliterate the forms or essences of the Animal or the Human whose roots go back to Aristotle, and whose classification into systems dates back to the eighteenth and nineteenth centuries (zoology for animals, hypothesis of polygenesis for human races). This hybridity would ensure a coevolution for the human species through technology and would include animals and all other living beings. This coevolution would radically modify the process of Darwinian evolution.

Such a conception of life rests on the hypothesis of John Von Neumann, who characterized life by its complexity and held that such complexity could be integrated into machines—artificial automatons.[42] The demarcation line of life forms is thus absorbed into the nonliving: there is thus nothing besides artificiality and hybridity. The entire field of synthetic biology, which is growing rapidly, for example, in the creation of artificial bacteria, is based on this hypothesis.

Is it necessary to naturalize monstrosity? Is it necessary to tame monstrous technologies? What value should be given to hybridity that establishes new forms of continuity between human beings, animals, and nature? Thinking about the status of the animal that has been modified by technology does not mean we can skirt our human responsibilities; the main moral consequence of such hybridity remains the need to consider the animal's own interests by rethinking our philosophical concept of nature.[43]

## CONCLUSION: THE ANIMAL AS A CHALLENGE TO MODERNITY

This overview of the animal question from an ethical, philosophical, and relational point of view clearly shows the new directions that Western philosophy is taking to rethink the relationship between humans and nature, starting with the status of the animal. One might ask: is this a new attempt to redefine the human in a modern environment that has been greatly modified by technology? Is this genuinely philosophical work that seeks to understand the animal and its world? By its proximity to the human, the animal has become a powerful reason to question the modern notion that separates humans from nature in some objectifying position, and that reduces animals to machines or units of production.

In the Anglo-Saxon world, and particularly North America, this questioning has been rooted in the primacy of ethics that, from Hume to contemporary utilitarianists, remains an ongoing preoccupation. The question of animal rights illustrates this primacy of ethics, and its normative force has changed many discourses and practices. The fact that this movement wrestles with conceptions of ethics largely derived from modernity—utilitarianism or Kantism—explains the strong critiques it has provoked from several environmental philosophers. Positing an essential difference between the wild and the domestic animal, environmental philosophers reaffirm their attachment to the concept of wild nature or wilderness that is rooted in a cultural and historical relationship between Americans and nature.

In the European, and more specifically Francophone, philosophical context, the questioning of the relationship between human beings and animals is possibly that much stronger because of the long legacy of Cartesianism that engraved the radical separation of humans from nature on many minds. The issue of animal rights constituted a strong catalyst for rethinking—from an ethical perspective—the cruel treatment inflicted on animals. Phenomenological studies have illustrated an immense richness for thinking the relationship with "the being of the animal." Drawing continuously on a philosophical tradition of twentieth-century thinkers—Husserl, Merleau-Ponty, Canguilhem, Simondon, Derrida, Deleuze—the phenomenological approach attempts a radical overhaul of the human relationship to animals and a strong attack on modernity. However, this approach cannot completely extricate itself from certain preconceptions about the human relationship to nature. These preconceptions are culturally constructed, rooted in the idea of making "good use" of a nature that has been landscaped and cultivated for a long time. The phenomenological approach nevertheless has the merit of opening our minds and sensibilities to discover "natural and artificial animal worlds" that will also, we can hope, deepen the understanding of our commonly shared environments transformed by modern technology and open up a new way of living together—dare we say—more harmoniously.

## NOTES

1. Catherine Larrère, "Des animaux-machines aux machines animales," in *Qui sont les animaux?*, ed. Jean Birnbaum (Paris: Folio, 2010), 88–109.
2. Translator's note: the French *homme* can be used to refer to both "man" and "human." It is being translated as "man" in this case to retain the emphasis on a phallocentric view of humans.
3. Georges Canguilhem, *Études d'histoire et de philosophie des sciences* (Paris: Vrin, 1987).
4. Georges Chapouthier, *Au bon vouloir de l'homme, l'animal* (Paris: Denoël, 1990).
5. It is worth remembering that Mill was deeply anti-naturalist: a duty does not follow logically from a fact. This position has important consequences for utilitarianism that does not hold to the idea of "obeying nature" since nature cannot constitute a moral foundation.
6. The debate between rule utilitarians and act utilitarians emerged in the 1960s with David Lyons (*Forms and Limits of Utilitarianism* [Oxford: Clarendon Press, 1965]) and was taken up again by J.J.C. Smart and Bernard Williams in *Utilitarianism: For and Against* (Cambridge: Cambridge University Press, 1973).

7. Carl Cohen and Tom Regan, *The Animal Rights Debate* (New York: Rowman & Littlefield Publishers, 2001); Cass Sunstein and Martha Nussbaum, eds., *Animal Rights: Current Debates and New Directions* (New York: Oxford University Press, 2004); Julian Franklin, *Animal Rights and Moral Philosophy* (New York: Columbia University Press, 2005).
8. Tom Regan and Peter Singer, *Animal Rights and Human Obligations* (Englewood Cliffs, NJ: Prentice Hall, 1989).
9. J. Baird Callicott, "Animal Liberation: A Triangular Affair," *Environmental Ethics* 2 (1980): 311–338.
10. Aldo Leopold, *A Sand County Almanac* (New York: Oxford University Press, 1949).
11. André Beauchamp, "L'animal dans la représentation chrétienne du monde," in *L'être humain, l'animal et la technique*, ed. Marie-Hélène Parizeau and Georges Chapouthier (Québec: Presses de l'Université Laval, 2007), 45–61.
12. Bryan Norton, *Toward Unity among Environmentalists* (New York: Oxford University Press, 1991); J. Baird Callicott and Michael Nelson, eds., *The Great New Wilderness Debate* (Athens: University of Georgia Press, 1998).
13. Roderick Nash, *Wilderness and the American Mind* (New Haven: Yale University Press, 1967).
14. Max Oelschlaeger, *The Idea of Wilderness: From Prehistory to the Age of Ecology* (New Haven: Yale University Press, 1993).
15. Georges Gusdorf, *Le savoir romantique de la nature* (Paris: Payot, 1985).
16. The definition of wilderness as found in the Wilderness Act is: "A wilderness, in contrast with those areas where man and his own works dominate the landscape, is hereby recognized as an area where the earth and community of life are untrammeled by man, where man himself is a visitor who does not remain."
17. Janine Chanteur, *Conclusion du Livre blanc sur l'expérimentation animale* (Paris: CNRS-Édition, 1995).
18. Sonia Desmoulins, "L'animal, objet d'invention brevetable," in *L'être humain*, ed. Parizeau and Chapouthier, 135–162.
19. Jean-Claude Nouët and Georges Chapouthier, eds., *Humanité, animalité: Quelles frontières?* (Paris: Édition Connaissances et Savoirs, 2006).
20. Georges Chapouthier, *Kant et le chimpanzé: Essai sur l'être humain, la morale et l'art* (Paris: Belin, 2009).
21. Joseph-Paul Beaufays, ed., "Les méthodes alternatives à l'expérimentation animale: Choix éthique et meilleure science?" *Revue des Questions Scientifiques* 176 (2005).
22. Joseph Duchêne, Joseph-Paul Beaufays, and Laurent Ravez, eds., *Entre l'homme et l'animal: Une nouvelle alliance?* (Namur: Les Presses de l'Université de Namur, 2002).
23. Commonly known as the 3Rs—replacement, reduction, refinement—this rule was stated in 1959 by British biologists William Russell and Rex Burch. It has, however, taken a long time for the rule to be adopted by scientists in their daily practice.
24. Joseph-Paul Beaufays, Laura Spano, Elisa Di Pietro, Bénédicte Schils, "La souris dans nos laboratoires: Entre anthropocentrisme et biocentrisme," *Revue des Questions Scientifiques* 176, no. 3–4 (2005): 323–329.
25. Jacques Derrida, *L'animal que donc je suis* (Paris: Galilée, 2006).
26. Élisabeth de Fontenay, *Sans offenser le genre humain: Réflections sur la cause animale* (Paris: Michel, 2008), 201–213.
27. De Fontenay, *Sans offenser*, 209.
28. Derrida, *L'animal*, 18.

29. This openness can be seen in Merleau-Ponty's description of studying the animal: "All zoology assumes from our side a methodical *Einfuehlung* [translator's note: empathy] into animal behaviour, with the participation of the animal in our perceptive life and the participation of our perceptive life in animality" (in *Themes from the Lectures at the College de France, 1952–1960*, trans. John O'Neill [Evanston, IL: Northwestern University Press, 1970], 97–8).
30. Georges Canguilhem, *La connaissance de la vie* (Paris: Vrin, 1985), 144–145.
31. Vinciane Despret, *Penser comme un rat* (Paris: Éditions Quae, 2009).
32. Dominique Lestel, *Les origines animales de la culture* (Paris: Flammarion, 2001).
33. Dominique Lestel, *L'animal singulier* (Paris: Seuil, 2004).
34. Joëlle Proust, *Comment l'esprit vient aux bêtes* (Paris: Gallimard, 1997).
35. Catherine Larrère and Raphaël Larrère, *Du bon usage de la nature* (Paris: Aubier, 1997).
36. Jean-Claude Gens, "Aux origines de l'éthique écologique: Jakob von Uexküll et Aldo Leopold en dialogue," in *Repenser la nature: Dialogue philosophique, Europe, Asie, Amériques*, ed. Jean-Philippe Pierron and Marie-Hélène Parizeau, 219–231 (Québec: Les Presses de l'Université Laval, 2012).
37. Augustin Berque, *Écoumène: Introduction à l'étude des milieux humains* (Paris: Belin, 2000); Augustin Berque, *Être humains sur la terre: Principes d'éthique de l'écoumène* (Paris: Gallimard, 1996).
38. Jürgen Habermas, *L'avenir de la nature humaine: Vers un eugénisme libéral?* (Paris: Gallimard, 2002).
39. Parizeau and Chapouthier, eds., *L'être humain.*
40. In her "Cyborg Manifesto," Donna Haraway describes the human-machine hybrid in the following way: "By the late twentieth century, our time, a mythic time, we are all chimeras, theorized and fabricated hybrids of machine and organism; in short, we are cyborgs. The cyborg is our ontology; it gives us our politics. The cyborg is a condensed image of both imagination and material reality, the two joined centers structuring any possibility of historical transformation" (in *Simians, Cyborgs, and Women: The Reinvention of Nature* [New York: Routledge, 1991], 149–181). Haraway's manifesto had a considerable impact on feminist and American intellectual circles because it dared to confront the issue of the technological transformation of human beings, and women in particular, as a possibility for going beyond the relationships of force, and for living in a diversity of almost unlimited forms of hybridity.
41. Marie-Hélène Parizeau, *Biotechnologies, nanotechnologies, écologie: Entre science et idéologie* (Paris: Éditions Quae, 2010).
42. John Von Neumann and Arthur Burks, *The Theory of Self-Reproducing Automata* (Champaign-Urbana: University of Illinois Press, 1966).
43. Pierron and Parizeau, eds., *Repenser la nature.*

# Bibliography


Abbot, Alison. "Regulations Proposed for Animal-Human Chimeras." *Nature* 475 (2011). Online at http://www.nature.com/news/2011/110721/full/475438a.html.

Adams, Carol. "Bestiality: The Unmentioned Abuse." *The Animals' Agenda* 15 (1995): 29–31.

Adams, Carol. *The Sexual Politics of Meat: A Feminist-Vegetarian Critical Theory*. New York: Continuum, 2000.

Afeissa, Hicham-Stéphane, and Jean-Baptiste Jeangène Vilmer, eds. *Philosophie animale: Différence, responsabilité, et communauté*. Paris: Vrin, 2010.

Agamben, Giorgio. *The Open: Man and Animal*. Translated by Kevin Attell. Palo Alto, CA: Stanford University Press, 2003.

Ageron, Charles-Robert. "L'Exposition Coloniale de 1931: Mythe républicain ou mythe impérial?" In *Les lieux de mémoire*, ed. Pierre Nora, vol. 1, *La République*, 561–591. Paris: Gallimard, 1984.

Agulhon, Maurice. "Le sang des bêtes: Le problème de la protection des animaux en France au XIXè siècle." In *Si les lions pouvaient parler: Essais sur la condition animale*, ed. Boris Cyrulnik, 1185–1219. Paris: Gallimard, 1998.

Aït-Touati, Frédérique. *Fictions of the Cosmos: Science and Literature in the Seventeenth Century*. Translated by Susan Emanuel. Chicago: University of Chicago Press, 2011.

Amiez, Céline, and Jean-Paul Joseph. "Rôle du cortex cingulaire antérieur dans les choix comportementaux basés sur les récompenses." In *Autour de l'éthologie et de la cognition animale*, ed. Fabienne Delfour and Michel Jean Dubois, 35–47. Lyon: Presses Universitaires de Lyon, 2005.

Armbruster, Karla, and Kathleen Wallace, eds. *Beyond Nature Writing: Expanding the Boundaries of Ecocriticism*. Charlottesville: University of Virginia Press, 2001.

Armengaud, Françoise. "L'anthropomorphisme: Vraie question ou faux débat?" In *Les animaux d'élevage ont-ils droit au bien-être?*, ed. Florence Burgat and Robert Dantzer, 203–231. Paris: INRA, 2001.

Aronson, Nicole. "'Que diable allait-il faire dans cette galère?': Mlle de Scudéry et les animaux." In *Les trois Scudéry*, ed. Alain Niderst, 523–532. Paris: Klincksieck, 1993.

Arthus-Bertrand, Yann, director. *Home*. Film. Beverly Hills, CA: Twentieth-Century Fox, 2009.

ASEF (Association Santé Environnement France). *Femmes enceintes et produits chimiques, la synthèse de l'ASEF*. Online at http://www.asef-asso.fr/mon-enfant/ma-grossesse/1255-femmes-enceintes-et-produits-chimiques-la-synthese-de-l-asef. (2012).

Ashworth, William, Jr. "Emblematic Natural History of the Renaissance." In *Cultures of Natural History*, ed. Nicolas Jardine and Emma Spary, 13–37. Cambridge: Cambridge University Press, 1996.

Atterton, Peter, and Matthew Calarco, eds. *Animal Philosophy: Essential Readings in Continental Thought*. New York: Continuum, 2004.

Austin, J. L. *Philosophical Papers*. Oxford: Oxford University Press, 1979.

AXA-Ipsos. *La perception individuelle des risques climatiques*. Online at http://www.axa.com/lib/axa/uploads/cahiersaxa/Etude-AXA-Ipsos_risques-climatiques.pdf. (2012).

Azouvi, François. *Descartes et la France: Histoire d'une passion nationale*. Paris: Fayard, 2002.

Azouvi, François. "Entre Descartes et Leibnitz: L'animisme dans les essais de physique de Claude Perrault." *Recherches sur le XVIIe siècle* 5 (1982): 9–19.

Backer, Dorothy Anne Liot. *Precious Women: A Feminist Phenomenon in the Age of Louis XIV.* New York: Basic Books, 1974.

Bailly, Jean-Christophe. "Les animaux conjuguent les verbes en silence." *L'Esprit créateur* 51 (2011): 106–114.

Bailly, Jean-Christophe. *Le dépaysement: Voyages en France.* Paris: Seuil, 2011.

Bailly, Jean-Christophe. *Le versant animal.* Paris: Bayard, 2007.

Baluška, František, Stefano Mancuso, Dieter Volkmann, and Peter Barlow. "Root Apices as Plant Command Centres: The Unique 'Brain-like' Status of the Root Apex Transition Zone." *Biologia* 59 (2004): 1–13.

Baratay, Éric. *Le point de vue animal: Une autre version de l'histoire.* Paris: Seuil, 2012.

Baratay, Éric. "Les socio-anthropo-logues et les animaux." *Sociétés* 108 (2010): 9–18.

Baratay, Éric. "Pour une histoire éthologique et une éthologie historique." *Études rurales* 189 (2012): 91–106.

Baratay, Éric, and Elisabeth Hardouin-Fugier. *Zoo: A History of Zoological Gardens in the West.* London: Reaktion Books, 2002.

Baratay, Éric, and Jean-Luc Mayaud. "Un champ pour l'histoire: L'animal." *Cahiers d'histoire* 42 (1997): 3–4, 409–442.

Baschet, René. "L'Exposition Coloniale." *L'Illustration.* Paris: L'Illustration, 1931.

Bates, Lucy, and Richard Byrnes. "Creative or Created: Using Anecdotes to Investigate Animal Cognition." *Methods* 42 (2007): 12–21.

Baudin, Matthieu. *Le développement durable: Nouvelle idéologie pour le XXIe siècle?* Paris: L'Harmattan, 2009.

Beauchamp, André. "L'animal dans la représentation chrétienne du monde." In *L'être humain, l'animal et la technique*, ed. Marie-Hélène Parizeau and Georges Chapouthier, 45–61. Québec: Presses de l'Université Laval, 2007.

Beaufays, Joseph-Paul, ed. "Les méthodes alternatives à l'expérimentation animale: Choix éthique et meilleure science?" *Revue des Questions Scientifiques* 176 (2005).

Beaufays, Joseph-Paul, Laura Spano, Elisa Di Pietro, and Bénédicte Schils. "La souris dans nos laboratoires: Entre anthropocentrisme et biocentrisme." *Revue des Questions Scientifiques* 176 (2005): 323–329.

Beck, Corinne, and Éric Fabre. "Interroger le loup historique? Entre la biologie et l'histoire: Un dialogue interdisciplinaire." In *Repenser le sauvage grâce au retour du loup*, ed. Jean-Marc Moriceau and Philippe Madeline, 13–21. Caen: Maison des Sciences Humaines, 2010.

Belon, Pierre. *Voyage au Levant.* Paris: Gilles Corrozet, 1553.

Bentham, Jeremy. *An Introduction to the Principles of Morals and Legislation.* Oxford, Clarendon Press, 1907.

Benyus, Janine. *Biomimicry: Innovation Inspired by Nature.* New York: William Morrow, 1997.

Berger, John. *About Looking.* New York: Pantheon, 1980.

Berger, John. *Ways of Seeing.* London: BBC Corporation, 1972.

Berque, Augustin. *Écoumène: Introduction à l'étude des milieux humains.* Paris: Belin, 2000.

Berque, Augustin. *Être humains sur la terre: Principes d'éthique de l'écoumène.* Paris: Gallimard, 1996.

Bhabha, Homi K. "The Other Question: Difference, Discrimination, and the Discourse of Colonialism." In *Black British Cultural Studies: A Reader*, ed. Houston A. Baker Jr., Manthia Diawara, and Ruth H. Lindeborg. Chicago: University of Chicago Press, 1996.

Bichet, Yves. *La part animale*. Paris: Gallimard, 1994.

Bird-David, Nurit. "Animism Revisited." *Current Anthropology* 40 (1999): 67–91.

Blake, William. *Songs of Experience*. New York: Dover, 1984.

Blanc, Nathalie. "Éthique et esthétique de l'environnement." *EspacesTemps.net*. Online at http://www.espacestemps.net/en/articles/thique-et-esthetique-de-lrsquoenvironnement-en. (2008).

Blanc, Nathalie. "La nature dans la cité." Doctoral thesis, Université de Paris 1 Sorbonne, 1996.

Blanc, Nathalie. *L'animal dans la ville*. Paris: Odile Jacob, 2000.

Blanc, Nathalie. "La place de l'animal dans les politiques urbaines." *Communications* 74 (2003) ("Bienfaisante nature"): 159–174.

Blanc, Nathalie, and Marianne Cohen. "L'animal: Une figure contemporaine de la géographie." *Espaces et sociétés* 110–111 (2003): 25–40.

Bloch, Marc. *Apologie pour l'histoire ou le métier d'historien*. Paris: Armand Colin, 1997.

Blondiaux, Loic, and Jean-Michel Fourniaux. "Un bilan des recherches sur la participation du public en démocratie: Beaucoup de bruit pour rien?" *Participations* 1 (2011): 8–35.

Bondolfi, Alberto, ed. *L'homme et l'animal: Dimensions éthiques de leur relation*. Fribourg: Éditions universitaires, 1995.

Bourg, Dominique, and Kerry Whiteside. *Vers une démocratie écologique: Le citoyen, le savant et le politique*. Paris: Seuil, 2010.

Bozonnet, Jean-Paul. "Les métamorphoses du grand récit écologiste et son appropriation par la société civile." *Revue d'Allemagne et des pays de langue allemande* 9 (2007): 311–342.

Brown, Harcourt. *Science and the Human Comedy: Natural Philosophy in French Literature from Rabelais to Maupertuis*. Toronto: University of Toronto Press, 1976.

Budiansky, Serge. *The Covenant of the Wild: Why Animals Choose Domestication*. New Haven: Yale University Press, 1999.

Buell, Lawrence. *Writing for an Endangered World: Literature, Culture, and Environment in the U.S. and Beyond*. Cambridge, MA: Belknap, 2001.

Burgat, Florence. *Animal mon prochain*. Paris: Odile Jacob, 1997.

Burgat, Florence. *Liberté et inquiétude de la vie animale*. Paris: Éditions Kimé, 2006.

Burgat, Florence, ed. *Penser le comportement animal*. Paris: Quae, 2010.

Burgat, Florence. *Une autre existence: La condition animale*. Paris, Albin Michel, 2012.

Burt, Jonathan. *Animals in Film*. London: Reaktion Books, 2002.

Buytendijk, Frederik. *L'homme et l'animal: Essai de psychologie comparée*. Translated by Rémi Laureillard. Paris: Gallimard, 1965.

Caballero, Francis. *Droit du sexe*. Paris: LGDJ, 2010.

Callicott, J. Baird. "Animal Liberation: A Triangular Affair." *Environmental Ethics* 2 (1980): 311–338.

Callicott, J. Baird. *In Defense of the Land Ethic*. New York: State University of New York Press, 1989.

Callicott, J. Baird, and Michael Nelson, eds. *The Great New Wilderness Debate*. Athens: University of Georgia Press, 1998.

Canguilhem, Georges. *Études d'histoire et de philosophie des sciences*. Paris: Vrin, 1987.

Canguilhem, Georges. *La connaissance de la vie*. Paris: Vrin, 1985.

Carroll, Lewis. *Alice's Adventures in Wonderland; and, Through the Looking-Glass: and What Alice Found There*. Edited by Hugh Haughton. London: Penguin, 1998.

Carson, Rachel. *Silent Spring*. 40th Anniversary Edition. 1962; New York: Houghton Mifflin, 2002.

Cavalieri, Paola, and Peter Singer, eds. *The Great Ape Project: Equality beyond Humanity*. New York: Saint Martin's Griffin, 1996.

Célestin, Roger, Éliane DalMolin, Anne Mairesse, and Anne Simon, eds. *Contemporary French and Francophone Studies (Sites)* 16 (2012).

Chacornac, Jérôme. "La définition sur mesure d'une infraction à la finalité incertaine." *Dalloz, Etudes et commentaires* 8 (2008): 524–527.

Chanteur, Janine. *Conclusion du Livre blanc sur l'expérimentation animale.* Paris: CNRS-Édition, 1995.

Chapouthier, Georges. *Au bon vouloir de l'homme, l'animal.* Paris: Denoël, 1990.

Chapouthier, Georges. *Kant et le chimpanzé: Essai sur l'être humain, la morale et l'art.* Paris: Belin, 2009.

Chapouthier, Georges, ed. *L'animal humain: Traits et spécificités.* Paris: L'Harmattan, 2004.

Chapouthier, Georges, and Jean-Claude Nouët, eds. *Les droits de l'animal aujourd'hui.* Paris: Arléa-Corlet-Ligue française des droits de l'animal, 1997.

Chapouthier, Georges, and Frédéric Kaplan. *L'homme, l'animal et la machine.* Paris: CNRS Editions, 2011.

Christen, Yves. *L'animal est-il une personne?* Paris: Flammarion, 2009.

Clark, David. "On Being 'The Last Kantian in Nazi Germany': Dwelling with Animals after Levinas." In *Animal Acts: Configuring the Human in Western History*, ed. Jennifer Ham and Matthew Senior, 165–198. New York: Routledge, 1997.

Cohen, Carl, and Tom Regan. *The Animal Rights Debate.* New York: Rowman & Littlefield, 2001.

Cole, Francis. *A History of Comparative Anatomy: From Aristotle to the Eighteenth Century.* London: Dover, 1975.

Coppinger, Raymond, and Lorna Coppinger. *Dogs: A New Understanding of Canine Origin, Behavior, and Evolution.* Chicago: University of Chicago Press, 2002.

Cosans, Christopher. "Aristotle's Anatomical Philosophy of Nature." *Biology and Philosophy* 13 (1998): 311–339.

Culler, Jonathan. "Philosophy and Literature: The Fortunes of the Performative." *Poetics Today* 21 (2000): 503–519.

Cusset, François. *French Theory: Foucault, Derrida, Deleuze & Cie et les mutations de la vie intellectuelle aux États-Unis.* Paris: La Découverte, 2003.

Cvetkovich, Ann. *An Archive of Feelings: Trauma, Sexuality, and Lesbian Public Cultures.* Durham, NC: Duke University Press, 2003.

Daston, Lorraine. "Classifications of Knowledge in the Age of Louis XIV." In *Sun King: The Ascendancy of French Culture during the Reign of Louis XIV*, ed. David Lee Rubin, 207–220. Washington, DC: Folger Books, 1997.

Daston, Lorraine, and Peter Galison. *Objectivity.* New York: Zone Books, 2007.

Dawkins, Richard. *The Selfish Gene.* Oxford: Oxford University Press, 1976.

Deleuze, Gilles, and Félix Guattari. *A Thousand Plateaus 2: Capitalism and Schizophrenia.* Translated by Brian Massumi. Minneapolis: University of Minnesota Press, 1987.

de Fontenay, Élisabeth. "La bête est sans raison." *Critique* 375–376 (1978): 707–729.

de Fontenay, Élisabeth. *Le silence des bêtes: La philosophie à l'épreuve de l'animalité.* Paris: Fayard, 1999.

de Fontenay, Élisabeth. *Without Offending Humans: A Critique of Animal Rights.* Translated by Will Bishop. Minneapolis: University of Minnesota Press, 2012.

DeJean, Joan. "Amazons and Literary Women: Female Culture during the Reign of the Sun King." In *Sun King: The Ascendancy of French Culture during the Reign of Louis XIV*, ed. David Lee Rubin, 115–128. Washington, DC: Folger Books, 1992.

Delannoy, Isabelle. "Sciences, humanités et écologie: Trouver les nouvelles passerelles et les nouveaux mots pour entreprendre un nouveau monde." *French Literary Studies* 39 (2012): 139–154.

Demaison, André. *Guide officiel à l'Exposition Coloniale Internationale.* Paris: Éditions Mayeux, 1931.

de Man, Paul. *Allegories of Reading: Figural Language in Rousseau, Nietzsche, Rilke, and Proust.* New Haven: Yale University Press, 1979.

de Man, Paul. *The Resistance to Theory.* Foreword by Wlad Godzich. Minneapolis: University of Minnesota Press, 1986.

de Man, Paul. "The Return to Philology." *Times Literary Supplement,* December 10, 1982.

Derrida, Jacques. *The Animal That Therefore I Am.* Edited by Marie-Louise Mallet. Translated by David Wills. New York: Fordham University Press, 2008.

Derrida, Jacques. "'Eating Well,' or the Calculation of the Subject: An Interview with Jacques Derrida." In *Who Comes after the Subject,* ed. Eduardo Cadava, Peter Connor, and Jean-Luc Nancy, 96–119. New York: Routledge, 1991.

Derrida, Jacques. *Of Grammatology.* Translated by Gayatri Chakravorty Spivak. Baltimore: Johns Hopkins University Press, 1976.

Derrida, Jacques. *Politics of Friendship.* Translated by George Collins. New York: Verso, 1997.

Derrida, Jacques. *Specters of Marx: The State of the Debt, the Work of Mourning, and the New International.* Translated by Peggy Kamuf. New York: Routledge, 2006.

Derrida, Jacques. *The Work of Mourning.* Edited by Pascale-Anne Brault and Michael Naas. Chicago: University of Chicago Press, 2001.

Derrida, Jacques. *Writing and Difference.* Translated by Alan Bass. Chicago: University of Chicago Press, 1978.

Derrida, Jacques, and Elisabeth Roudinesco. *For What Tomorrow: A Dialogue.* Translated by Jeff Fort. Stanford, CA: Stanford University Press, 2004.

Desblache, Lucile. *Bestiaire du roman contemporain d'expression française.* Clermont-Ferrand: Presses de l'Université Blaise Pascal, 2002.

Desblache, Lucile. *La plume des bêtes: Les animaux dans le roman.* Paris: L'Harmattan, 2011.

Desblache, Lucile, ed. *Hybrides et monstres: Transgressions et promesses des cultures contemporaines.* Dijon: Presses Universitaires de Dijon, 2012.

Descartes, René. *Discours de la méthode.* Leiden: Maire, 1637.

Descartes, René. *Lettres de Mr. Descartes, où sont traitées plusieurs belles questions touchant la morale, la physique, la médecine, et les mathématiques.* Paris: Angot, 1657.

Des Chene, Dennis. "Mechanisms of Life in the Seventeenth Century: Borelli, Perrault, Regis." *Studies in the History and Philosophy of Biology and the Biomedical Sciences* 36 (2005): 245–260.

Descola, Philippe. *Beyond Nature and Culture.* Translated by Janet Lloyd. Foreword by Marshall Sahlins. Chicago: University of Chicago Press, 2013.

Desmoulins, Sonia. "L'animal, objet d'invention brevetable." In *L'être humain, l'animal et la technique,* ed. Marie-Hélène Parizeau and Georges Chapouthier, 135–162. Québec: Presses de l'Université Laval, 2007.

Despret, Vinciane. "Culture and Gender Do Not Dissolve into How Scientists Read the World." In *Rebels, Mavericks, and Heretics in Biology,* ed. Oren Harman and Michael Dietrich, 338–355. New Haven: Yale University Press, 2008.

Despret, Vinciane. "Ecology and Ideology: The Case of Ethology." *International Problems* 63 (1994): 45–61.

Despret, Vinciane. *Naissance d'une théorie éthologique: La danse du cratérope écaillé.* Paris: Les Empêcheurs de penser en rond, 1996.

Despret, Vinciane. *Our Emotional Makeup: Ethnopsychology and Selfhood.* Translated by Marjolijn de Jager. New York: Other Press, 2004.

Despret, Vinciane. *Penser comme un rat.* Paris: Quae, 2009.

Despret, Vinciane. *Quand le loup habitera avec l'agneau.* Paris: Les Empêcheurs de penser en rond, 2002.

Despret, Vinciane. *Que diraient les animaux, si . . . on leur posait les bonnes questions?* Paris: La Découverte, 2012.

Despret, Vinciane. "Rencontrer, avec Donna Haraway un animal." *Critique* 747–8 (2009): 747–759.

Despret, Vinciane. "Sheep Do Have Opinions." In *Making Things Public: Atmospheres of Democracy*, ed. Bruno Latour and Peter Weibel, 360–370. Cambridge, MA: MIT Press, 2006.

Despret, Vinciane, and Serge Gutwith. "L'affaire Harry: Petite scientification." *Terrain* 52 (2009): 142–152.

Despret, Vinciane, and Jocelyne Porcher. *Être bête.* Arles: Actes Sud, 2007.

Despret, Vinciane, and Isabelle Stengers. *Les faiseuses d'histoires: Que font les femmes à la pensée?* Paris: Les Empêcheurs de penser en rond/La Découverte, 2011.

Détrez, Christine, and Anne Simon. *À leur corps défendant: Les femmes à l'épreuve du nouvel ordre moral.* Paris: Seuil, 2006.

de Waal, Frans. *The Ape and the Sushi Master.* New York: Basic Books, 2002.

Dewitte, Jacques. "L'anthropomorphisme, voie d'accès privilégiée au vivant: L'apport de Hans Jonas." *Revue philosophique de Louvain* 100 (2002): 437–465.

Duchêne, Joseph, Joseph-Paul Beaufays, and Laurent Ravez, eds. *Entre l'homme et l'animal: Une nouvelle alliance?* Namur, Belgium: Presses Universitaires de Namur, 2002.

Dufour-Maître, Myriam. *Les précieuses: Naissance des femmes de lettres au XVIIe siècle.* Paris: Honoré Champion, 2008.

Dunayer, Joan. "Sexist Words, Speciesist Roots." In *Animals and Women: Feminist Theoretical Explorations*, ed. Carol J. Adams and Josephine Donovan, 11–26. Durham, NC: Duke University Press, 1995.

Dunlap, Riley, Kent Van Liere, Angela Mertig, and Robert Emmet Jones. "Measuring Endorsement of the New Ecological Paradigm: A Revisited NEP Scale." *Journal of Social Issues* 56 (2000): 425–442.

Engélibert, Jean-Paul, Lucie Campos, Catherine Coquio, and Georges Chapouthier, eds. *La question animale: Entre science, littérature et philosophie.* Rennes: Presses Universitaires de Rennes, 2011.

Fabre, Éric, and Julien Alleau. "La disparition des loups ou essai d'écologie historique." In *L'animal sauvage entre nuisance et patrimoine*, ed. Stéphane Frioux and Émilie Pépy, 25–34. Lyon: ENS, 2009.

Ferry, Luc. *Le nouvel ordre écologique: L'arbre, l'animal, et l'homme.* Paris: Grasset, 1992.

Finnemore, Martha. *The Purpose of Intervention: Changing Beliefs about the Use of Force.* Ithaca, NY: Cornell University Press, 2003.

Fontenelle, Bernard de. *Histoire de l'Académie royale des sciences, depuis son établissement en 1666 jusqu'à 1686.* 10 vols. Paris: Martin, Coignard, Guérin, 1733.

Foucart-Walter, Elisabeth. *Pieter Boel, 1622–1674: Peintre des animaux de Louis XIV. Le fonds des études peintes des Gobelins.* Paris: Réunion des musées nationaux, 2001.

Foucault, Michel. *The Order of Things: An Archeology of the Human Sciences.* New York: Pantheon, 1994.

Franklin, Julian. *Animal Rights and Moral Philosophy.* New York: Columbia University Press, 2005.

Freccero, Carla. *Queer/Early/Modern.* Durham, NC: Duke University Press, 2005.

Ganim, Russell. "Scientific Verses: Subversion of Cartesian Theory and Practice in the 'Discours à Madame de La Sablière.'" In *Refiguring La Fontaine: Tercentenary Essays*, ed. Anne L. Birberick, 101–111. Charlottesville, VA: Rookwood Press, 1996.

Génetiot, Alain. *Le classicisme.* Paris: Presses Universitaires de France, 2005.

Gens, Jean-Claude. "Aux origines de l'éthique écologique: Jakob von Uexküll et Aldo Leopold en dialogue." In *Repenser la nature: Dialogue philosophique, Europe, Asie, Amériques*, ed. Jean-Philippe Pierron and Marie-Hélène Parizeau, 219–231. Québec: Les Presses de l'Université Laval, 2012.

George, Wilma. *Animals and Maps*. London: Secker and Warburg, 1969.

Germain-Sée, Emile. *L'Animal dans la société devant la science, la philosophie, la religion*. Biarritz: Imprimerie moderne de "La Gazette," 1936.

Ginzburg, Carlo. "Signes, traces, pistes: Racines d'un paradigme de l'indice." *Le Débat* 6 (1980): 3–44.

Girardet, Raoul. *L'idée coloniale en France de 1871 à 1962*. Paris: Pluriel, La Table Ronde, 1972.

Godelier, Maurice. *The Mental and the Material*. Translated by Martin Thom. London: Verso, 1986.

Goffi, Jean-Yves. *Le philosophe et ses animaux*. Paris: Jacqueline Chambon, 1993.

Goffi, Jean-Yves. *Qu'est-ce que l'animalité?* Paris: Vrin, 2004.

Gompertz, Lewis. *Moral Inquiries on the Situation of Man and of Brutes*. New York: Centaur Press, 1992.

Goux, Jean-Joseph. "The Phallus: Masculine Identity and the 'Exchange of Women.'" *differences* 4 (1992): 40–75.

Greenberg, Sean. "Descartes on the Passions: Function, Representation, and Motivation." *Noûs* 41 (2007): 714–734.

Greenblatt, Stephen. *Marvelous Possessions: The Wonder of the New World*. Chicago: University of Chicago Press, 1991.

Grundmann, Emmanuelle. *Ces forêts qu'on assassine*. Paris: Calmann-Levy, 2007.

Guerrini, Anita. "The King's Animals and the King's Books: The Illustrations for the Paris Academy's *Histoire des Animaux*." *Annals of Science* 67 (2010): 383–404.

Guerrini, Anita. "Perrault, Buffon, and the History of Animals." *Notes and Records of the Royal Society*. Online at doi: 10.1098/rsnr.2012.0044. (2012).

Guerrini, Anita. "The 'Virtual Menagerie': The *Histoire des Animaux* Project." *Configurations* 14 (2006): 29–41.

Gusdorf, Georges. *Le savoir romantique de la nature*. Paris: Payot, 1985.

Habermas, Jürgen. *The Future of Human Nature*. Cambridge: Polity, 2003.

Hache, Emilie. *Ce à quoi nous tenons*. Paris: La Découverte, 2011.

Hahn, Roger. *The Anatomy of a Scientific Institution: The Paris Academy of Sciences, 1666–1803*. Berkeley: University of California Press, 1971.

Hallé, Francis. *In Praise of Plants*. Translated by David Lee. Portland, OR: Timber Press, 2011.

Hallé, Francis. *Plaidoyer pour l'arbre*. Paris: Actes Sud, 2005.

Hallowell, Alfred Irving. "Ojibwa Ontology, Behavior, and Worldview." In *Culture in History: Essay in Honor of Paul Radin*, ed. Stanley Diamond, 17–49. New York: Octagon Books, 1960.

Haraway, Donna. *Primate Visions: Gender, Race, and Nature in the World of Modern Science*. New York: Routledge, 1989.

Haraway, Donna. *Simians, Cyborgs, and Women. The Reinvention of Nature*. New York: Routledge, 1991.

Haraway, Donna. *When Species Meet*. Minneapolis: University of Minnesota Press, 2008.

Harth, Erica. *Cartesian Women: Versions and Subversions of Rational Discourse in the Old Regime*. Ithaca, NY: Cornell University Press, 1992.

Harth, Erica. "Classical Science: *Mémoires pour servir à l'histoire naturelle des animaux*." In *Actes de Baton Rouge*, ed. Selma Zebouni, 208–217. Paris: Biblio 17, 1986.

Harvey, Graham. "Guesthood as Ethical Decolonising Research Method." *Numen* 50 (2003): 125–146.

Hayek, Friedrich August. *Droit, législation et liberté*. Foreword by Philippe Nemo. Paris: PUF, 2007.

Hearne, Vicki. *Adam's Task: Calling Animals by Name*. New York: Knopf/Random House, 1986.

Hegel, Georg Wilhelm Friedrich. *Encyclopedia of Philosophy*. Translated by Gustav Emil Mueller. New York: Philosophical Library, 1959.

Hernandez, Ludovico. *Les procès de bestialité au XVIe et XVIIe siècles: Documents judiciaires inédits publiés avec un avant-propos*. Paris: ARCANES, 1955.

Herzfeld, Chris. *Wattana: Un Orang-Outang à Paris*. Paris: Payot, 2012.

Hodeir, Catherine, and Michel Pierre. *L'Exposition coloniale: 1931*. Paris: Éditions Complexe, 1991.

Hoffman, Paul. "Modèle mécaniste et modèle animiste: De quelques aspects de la représentation du vivant chez Descartes, Borelli et Stahl." *Revue des sciences humaines* 59 (1992): 199–211.

Husserl, Edmund. *Cartesian Meditations: An Introduction to Phenomenology*. Translated by Dorion Cairns. The Hague: Nijhoff, 1960.

Iacub, Marcela. *Confessions d'une mangeuse de viande: Pourquoi je ne suis plus carnivore*. Paris: Fayard, 2011.

Iacub, Marcela. *De la pornographie en Amérique: La liberté d'expression à l'âge des démocraties délibératives*. Paris: Fayard, 2010.

Iacub, Marcela. *Le crime était presque sexuel et autres essais de casuistique juridique*. Paris: Champs Flammarion, 2003.

Iacub, Marcela. *Par le trou de la serrure: Une histoire de la pudeur publique*. Paris: Fayard, 2008.

Iacub, Marcela, and Patrice Maniglier. *Antimanuel d'éducation sexuelle*. Paris: Bréal, 2005.

Ingold, Tim. *Companion Encyclopedia of Anthropology*. London: Routledge, 2002.

Ingold, Tim. *Lines: A Brief History*. London: Routledge, 2007.

Ingold, Tim. "Point, Line and Counterpoint: From Environment to Fluid Space." In *Neurobiology of 'Umwelt': How Living Beings Perceive the World*, ed. Alain Berthoz and Yves Christen, 141–155. Berlin: Springer-Verlag, 2009.

Irigaray, Luce. *This Sex Which Is Not One*. Translated by Catherine Porter and Carolyn Burke. New York: Cornell University Press, 1985.

James, William. *A Pluralistic Universe*. London: Longmans, Green and Co., 1906.

James, William. "Pragmatism's Conception of Truth." In *Pragmatism: A New Name for Some Old Ways of Thinking*, 197–236. London: Longman, 1907.

Jonas, Hans. *The Imperative of Responsibility: In Search of an Ethics for the Technological Age*. Chicago: University of Chicago Press, 1984.

Jullien, François. *Un sage est sans idée*. Paris: Seuil, 1998.

Kafka, Franz. "A Report to an Academy." In *The Complete Stories and Parables*, ed. Nahum N. Glatzer. New York: Quality Paperback Book Club, 1988.

Kalaora, Bernard. "Naissance et développement d'un loisir urbain: La forêt de Fontainebleau." *Études rurales* 83 (1981): 97–115.

Keck, Frédéric, and Noélie Vialles, eds. *Des hommes malades des animaux*. Paris: L'Herne, 2012.

Keeley, Brian. "Anthropomorphism, primatomorphism, mammalomorphism: Understanding cross-species comparisons." *Biology and Philosophy* 19 (2004): 521–540.

Kendall, Karalyn. "The Face of a Dog: Levinasian Ethics and Human/Dog Coevolution." In *Queering the Non-Human*, ed. Noreen Giffney and Myra J. Hird, 185–204. Hampshire, UK: Ashgate, 2008.

Kofman, Sarah. *Autobiogriffures du chat Murr d'Hoffmann*. Paris: Galilée, 1984.

Kohler, Florent, ed. "Sociabilités animales." *Études rurales* 189, no. 1 (2012).

La Fontaine, Jean de. *Fables*. Paris: Garnier, 1923.

Laissus, Yves, and Jean-Jacques Petter. *Les animaux du Muséum, 1793–1993*. Paris: Muséum National d'Histoire Naturelle, 1993.

Laland, Kevin, and Bennet Galef. *The Question of Animal Culture*. Cambridge, MA: Harvard University Press, 2009.

Larrère, Catherine. "Care et environnement: La montagne ou le jardin?" In *Tous vulnérables? Le care, les animaux et l'environnement*, ed. Sandra Laugier, 233–263. Paris: Payot, 2012.

Larrère, Catherine. "Des animaux-machines aux machines animales." In *Qui sont les animaux?*, ed. Jean Birnbaum, 88–109. Paris: Folio, 2010.

Larrère, Catherine. "La naturalisation des artifices." In *L'être human, l'animal et la technique*, ed. Marie-Hélène Parizeau and Georges Chapouthier. 79–96. Quebec: Presses de l'Université Laval, 2007.

Larrère, Catherine, and Raphaël Larrère. *Du bon usage de la nature: Pour une philosophie de l'environnement.* Paris: Aubier, 1997.

Lasserre, Audrey, and Anne Simon. *Nomadismes des romancières contemporaines de langue française.* Paris: Presses de la Sorbonne nouvelle, 2008.

Latour, Bruno. "A Well-Articulated Primatology: Reflections of a Fellow Traveler." In *Primate Encounters: Models of Science, Gender, and Society*, ed. Shirley Strum and Linda Fedigan, 358–382. Chicago: University of Chicago Press, 2000.

Latour, Bruno. *We Have Never Been Modern.* Translated by Catherine Porter. Cambridge, MA: Harvard University Press, 1993.

Latour, Bruno, and Steve Woolgar. *Laboratory Life: The Social Construction of Scientific Facts.* Beverly Hills, CA: Sage Publications, 1979.

Lawrence, J. *The Horse in All His Varieties and Uses.* London: Arnold, 1829.

Leopold, Aldo. *A Sand County Almanac.* New York: Oxford University Press, 1949.

Lerch, A., P. Roy, F. Pachet, and L. Nagle. "Closed-loop Bird-computer Interactions: A New Method to Study the Role of Bird Calls." *Animal Cognition* 14 (2011): 203–211.

Lestel, Dominique. *Apologie du carnivore.* Paris: Fayard, 2011.

Lestel, Dominique. "Could Beethoven Have Been a Bird and Could Picasso Have Been a Fish? Philosophical Problems of an Ethology of Art." In *Logic and Sensibility*, ed. Shigaru Watanabe. Tokyo: Keio University Press, 2012.

Lestel, Dominique. *L'animal est l'avenir de l'homme.* Paris: Fayard, 2010.

Lestel, Dominique. *L'animal singulier.* Paris: Seuil, 2004.

Lestel, Dominique. *Les amis de mes amis.* Paris: Seuil, 2007.

Lestel, Dominique. *Les origines animales de la culture.* Paris: Flammarion, 2001.

Lestel, Dominique. "Non-Human Artistic Practices: A Challenge to the Social Sciences of the Future." *Social Sciences Information* 50 (2011): 3–4, 505–512.

Lestel, Dominique. "What Capabilities for the Animal?" *Biosemiotics* 4 (2011): 83–102.

Lestel, Dominique. "What Does It Mean to Observe Rationality?" In *Rational Animals, Irrational Humans*, ed. Shigeru Watanabe et al., 44–66. Tokyo: Keio University Press, 2009.

Lévi-Strauss, Claude. *Totemism.* Translated by Rodney Needham. London: Merlin Press, 1962.

Lévi-Strauss, Claude. *Tristes Tropiques.* Translated by John Russell. New York: Atheneum, 1965.

Licoppe, Christian. "The Crystallization of a New Narrative Form of Experimental Reports (1660–1690)." *Science in Context* 7 (1994): 205–244.

Lietaer, Bernard. *The Future of Money: A New Way to Create Wealth, Work and a Wiser World.* London: Century, 2001.

Llored, Patrick. *Jacques Derrida: Politique et éthique de l'animalité.* Mons: Sils Maria, 2013.

Loevenbruck, Pierre. *Animaux captifs: La vie des zoos.* Paris: La Toison d'or, 1954.

Lopez Espinosa, Maria José. "Maternal-child Exposure via the Placenta to Environmental Chemical Substances with Hormonal Activity." PhD diss., University of Grenada, Spain, 2007.

Lougee, Carolyn. *Le Paradis des femmes: Women, Salons, and Social Stratification in Seventeenth-Century France.* Princeton, NJ: Princeton University Press, 1976.

Lyons, David. *Forms and Limits of Utilitarianism.* Oxford: Clarendon Press, 1965.

Macé, Marielle. "Styles animaux." *L'Esprit créateur* 51 (2011): 97–105.

Magel, Charles. *Keyguide to Information Sources in Animal Rights*. Jefferson, NC: McFarland, 1989.

Marchitello, Howard. *The Machine in the Text: Science and Literature in the Age of Shakespeare and Galileo*. New York: Oxford University Press, 2011.

Maréchal, Jean-Yves. "Sévices graves ou actes de cruauté envers les animaux." *JurisClasseur Pénal*, articles 521–1 and 521–2: fascicule 10. Paris: LexisNexis SA, 2009.

Marguénaud, Jean-Pierre. *L'animal en droit privé*. Paris: PUF, 1992.

Margulis, Lynn, and Dorion Sagan. *Slanted Truths: Essays on Gaia, Symbiosis, and Evolution*. New York: Springer, 1997.

Mauss, Marcel. "Essai sur le don: Forme et raison de l'échange dans les sociétés archaïques." *Année Sociologique* 1 (1925).

McClaughlin, Trevor. "Censorship and Defenders of the Cartesian Faith in Mid-Seventeenth-Century France." *Journal of the History of Ideas* 40 (1979): 563–581.

Merleau-Ponty, Maurice. *La Nature: Notes—Cours du Collège de France*. Edited by Dominique Séglard. Paris: Seuil, 1995.

Merleau-Ponty, Maurice. *The Structure of Behavior*. Translated by A. L. Fisher. Boston: Beacon Press, 1963.

Merleau-Ponty, Maurice. *Themes from the lectures at the Collège de France, 1952–1960*. Translated by John O'Neill. Evanston, IL: Northwestern University Press, 1970.

Merleau-Ponty, Maurice. *The Visible and the Invisible*. Translated by Alfonso Lingis. Edited by Claude Lefort. Evanston, IL: Northwestern University Press, 1968.

Métherie, Jean-Claude de la. *De l'homme considéré moralement; de ses mœurs, et de celles des animaux*. Paris: Maradan, 1802.

Métherie, Jean-Claude de la. *Principes de la philosophie naturelle dans lesquels on cherche à déterminer les degrés de certitude et de probabilité des connaissances humaines*. Geneva: N.p., 1787.

Midgley, Mary. *Animals and Why They Matter*. Athens: University of Georgia Press, 1983.

Miklosi, Adam. *Dog Behaviour, Evolution and Cognition*. Oxford: Oxford University Press, 2008.

Mongrédien, Georges. *Les précieux et les précieuses*. Paris: Mercure de France, 1963.

Montaigne, Michel de. *The Complete Essays of Montaigne*. Translated by Donald Frame. Stanford, CA: Stanford University Press, 1957.

Moriceau, Jean-Marc. *Histoire du méchant loup*. Paris: Fayard, 2007.

Morton, Patricia. *Hybrid Modernities: Architecture and Representation at the 1931 Colonial Exposition, Paris*. Cambridge, MA: MIT Press, 2000.

Mullan, Bob, and Garry Marvin. *Zoo Culture*. London: Weidenfeld & Nicolson, 1987.

Nagajawa, Naofumi, Masayuki Nakarichi, and Hideki Sugiura. *The Japanese Macaques*. Tokyo: Springer, 2010.

Nash, Roderick. *Wilderness and the American Mind*. New Haven: Yale University Press, 1967.

Nathan, Debbie, and Michael Snedeker. *Satan's Silence: Ritual Abuse and the Making of a Modern American Witch Hunt*. New York: Author Choice Press, 2001.

Niderst, Alain. *Madeleine de Scudéry, Paul Pellisson et leur monde*. Paris: Presses Universitaires de France, 1976.

Nomadéis, K-Minos, and Semiocast. *La perception internationale du discours scientifique sur la menace climatique par le grand public dans six pays: Afrique du Sud, Brésil, Chine, États-Unis, France, Inde*. Report for the Centre d'analyse stratégique, October 2012.

Norton, Bryan. *Toward Unity among Environmentalists*. New York: Oxford University Press, 1991.

Nouët, Jean-Claude, and Georges Chapouthier, eds. *Humanité, animalité: Quelles frontières?* Paris: Connaissances et Savoirs, 2006.

Oelschlaeger, Max. *The Idea of Wilderness: From Prehistory to the Age of Ecology*. New Haven: Yale University Press, 1993.

Ostrom, Elinor. *Gouvernance des biens communs: Pour une approche nouvelle des ressources naturelles*. Bruxelles: De Boeck, 2011.

Parizeau, Marie-Hélène. *Biotechnologies, nanotechnologies, écologie: Entre science et idéologie*. Paris: Quae, 2010.

Pastoureau, Michel. *Les animaux célèbres*. Paris: Bonneton, 2001.

Perrault, Claude, compiler. *Description anatomique d'un caméléon, d'un castor, d'un dromadaire, d'un ours, et d'une gazelle*. Paris: Frederic Leonard, 1669.

Perrault, Claude, compiler. *Mémoires pour servir à l'histoire naturelle des animaux*. Paris: Imprimerie Royale, 1671.

Perrault, Claude, compiler. *Mémoires pour servir à l'histoire naturelle des animaux*. Paris: Imprimerie Royale, 1676.

Petitier, Paule, ed. *L'animal du XIXe siècle*. Online at http://www.equipe19.univ-paris-diderot.fr/Colloque animal/Page titre colloque.htm. (2008).

Picon, Antoine. *Claude Perrault, 1613–1688, ou La curiosité d'un classique*. Paris: Picard, 1988.

Pieterse, Jan Nederveen. *White on Black: Images of Africa and Blacks in Western Popular Culture*. New Haven: Yale University Press, 1992.

Plante et Cité. "Bienfaits du végétal en ville sur le bien-être et la santé humaine." Online at http://vegepolys.eu/media/rpc__n_special_bienfaits_nature_en_ville__015515700_1226_18122009.pdf. (2009).

Plumwood, Val. "Nature in the Active Voice." *Australian Humanities Review* 46 (2009): 113–129.

Poirier, Jacques. *L'animal littéraire: Des animaux et des mots*. Dijon: Presses Universitaires de Dijon, 2010.

Porcher, Jocelyne. *Bien-être animal et travail en élevage: Textes à l'appui*. Paris: INRA, 2004.

Porcher, Jocelyne. *Éleveurs et animaux, réinventer le lien*. Preface by Boris Cyrulnik. Paris: PUF, 2002.

Porcher, Jocelyne. "The Relationship between Workers and Animals in the Pork Industry: A Shared Suffering." *Journal of Agricultural and Environmental Ethics* 24 (2011): 3–17.

Porcher, Jocelyne. *Vivre avec les animaux: Une utopie pour le XXIe siècle*. Preface by Alain Caillé. Paris: Découverte, 2011.

Posthumus, Stephanie. "La nature et l'écologie chez Lévi-Strauss, Tournier, Serres." PhD diss., University of Western Ontario, 2003.

Posthumus, Stephanie. "Translating Ecocriticism: Dialoguing with Michel Serres." *Reconstruction: Studies in Contemporary Culture* 7 (2007). Online at http://reconstruction.eserver.org/072/posthumus.shtml.

Postdam Institute for Climate Impact Research and Climate Analytics. *Turn Down the Heat: Why a 4°C Warmer World Must Be Avoided*. Report for the World Bank, November 2012.

Proust, Joëlle. *Comment l'esprit vient aux bêtes*. Paris: Gallimard, 1997.

Putnam, Walter. "Please Don't Feed the Natives: Human Zoos, Colonial Desire, and Bodies on Display." *French Literature Series* 39 (2013): 55–68.

Ray, Paul, and Sherry Anderson. *The Cultural Creatives: How 50 Million People Are Changing the World*. New York: Harmony Books, 2000.

Redfield, Marc. "De Man, Schiller, and the Politics of Reception." *Diacritics* 20 (1990): 50–70.

Regan, Tom, and Peter Singer. *Animal Rights and Human Obligations*. Englewood Cliffs, NJ: Prentice Hall, 1989.

Richard, Jean-Pierre. *Proust et le monde sensible*. Paris: Seuil, 1974.

Rifkin, Jeremy. *The Empathic Civilization: The Race to Global Consciousness in a World in Crisis.* New York: Penguin, 2009.

Rifkin, Jeremy. *The Third Industrial Revolution: How Lateral Power Is Transforming Energy, the Economy, and the World.* New York: Palgrave Macmillan, 2011.

Ronell, Avital. *American philo: Entretiens avec Anne Dufourmantelle.* Paris: Stock, 2006.

Rosenfield, Leonora. *From Beast-Machine to Man-Machine: Animal Soul in French Letters from Descartes to La Mettrie.* New York: Oxford University Press, 1941.

Rothfels, Nigel. *Savages and Beasts: The Birth of the Modern Zoo.* Baltimore: Johns Hopkins University Press, 2002.

Rousseau, Jean-Jacques. *Discours sur l'origine et les fondements de l'inégalité parmi les hommes.* Paris: Aubier, 1973.

Sahlins, Peter. "The Royal Ménageries of Louis XIV and the Civilizing Process Revisited." *French Historical Studies* 35 (2012): 237–267.

Schaeffer, Jean-Marie. *La fin de l'exception humaine.* Paris: Gallimard, 2007.

Schapp, Wilhelm. *Empêtrés dans des histoires: L'être de l'homme et de la chose.* Translated by Jean Greisch. Paris: Éditions du Cerf, 1992.

Schmitt, Carl. *La notion de politique: Théorie du partisan.* Paris: Flammarion, 1972.

Schnapper, Antoine. *Le géant, la licorne, la tulipe: Collections françaises au XVIIe siècle.* Paris: Flammarion, 1988.

Schoentjes, Pierre. "Textes de la nature et nature du texte." *Poétique* 164 (2010): 477–479.

Schuiling, Jacqueline, and Wytze van der Naald. "A Present for Life: Hazardous Chemicals in Umbilical Cord Blood." Greenpeace-UK, 2005.

Scudéry, Madeleine de. "Histoire de deux Chaméléons." In *Nouvelles conversations de morale dédiées au Roy*, vol. 2: 496–629. Paris: La veuve de Sébastien Marbre-Cramoisy, 1688.

Searle, John. "Literal Meaning." *Erkenntnis* 13 (1978): 207–224.

Serres, Michel. *The Natural Contract.* Translated by Elizabeth MacArthur and William Paulson. Ann Arbor: University of Michigan Press, 1995.

Serres, Michel. "Revisiting the Natural Contract." Translated by Anne-Marie Feenberg-Dibon. Online at C-theory.net: 1000 Days of Theory, td039. (2006).

Shepard, Paul. *Thinking Animals.* Athens: University of Georgia Press, 1978.

Simon, Anne. "Hommes et bêtes à vif: Trouble dans la domestication et littérature contemporaine." In *Le moment du vivant*, ed. Arnaud François and Frédéric Worms. Paris: Presses Universitaires de France, 2013.

Singer, Peter. *Animal Liberation.* London: Cape, 1976.

Singer, Peter. "Heavy Petting." *Nerve* (March/April 2001).

Sloterdijk, Peter. *Essai d'intoxication volontaire.* Paris: Hachette, 2006.

Smart, J.J.C., and Bernard Williams. *Utilitarianism: For and Against.* Cambridge: Cambridge University Press, 1973.

Smuts, Barbara. "Reflections." In J. M. Coetzee, *The Lives of Animals*, 107–120. Princeton, NJ: Princeton University Press, 1999.

Smuts, Barbara, and John Watanabe. "Social Relationship and Ritualized Greetings in Adult Male Baboons (*papio cynocephalus Anubis*)." *International Journal of Primatology* 11 (1990): 147–172.

Soper, Kate. *What Is Nature? Culture, Politics, and the Non-Human.* Oxford: Blackwell, 1998.

Soury, Jules. "Anatomie et vivisection d'un caméléon." *Revue scientifique* 9 (1898): 336–337.

Specq, François. "Henry D. Thoreau et la naissance de l'idée de parc national." *Écologie & politique* 36 (2008): 29–40.

Sperber, Dan. "Pourquoi les animaux parfaits, les hybrides et les monstres sont-ils bons à penser symboliquement?" *L'Homme* 15 (1985): 5–34.

Spiegel, Marjorie. *The Dreaded Comparison: Human and Animal Slavery.* New York: Mirror Books, 1996.

Stengers, Isabelle. *Cosmopolitics 1: The Science Wars.* Translated by Robert Bononno. Minneapolis: University of Minnesota Press, 2010.

Stengers, Isabelle. *The Invention of Modern Science.* Translated by Daniel Smith. Minneapolis: University of Minnesota Press, 2000.

Stengers, Isabelle. *La vierge et le neutrino: Les scientifiques dans la tourmente.* Paris: Les Empêcheurs de penser en rond, 2006.

Stengers, Isabelle. *Thinking with Whitehead: A Free and Wild Creation of Concepts.* Translated by Michael Chase. Foreword by Bruno Latour. Cambridge, MA: Harvard University Press, 2011.

Strum, Shirley, and Linda Fedigan. "Changing Views of Primate Society: A Situated North American View." In *Primate Encounters: Models of Science, Gender, and Society*, ed. Shirley Strum and Linda Fedigan, 3–49. Chicago: University of Chicago Press, 2000.

Suberchicot, Alain. *Littérature et environnement: Pour une écocritique comparée.* Paris: Champion, 2012.

Sunstein, Cass, and Martha Nussbaum, eds. *Animal Rights: Current Debates and New Directions.* New York: Oxford University Press, 2004.

Sutton, G. V. *Science for a Polite Society: Gender, Culture, and the Demonstration of Enlightenment.* Boulder: University of Colorado Press, 1995.

Talon-Hugon, Carole. *Descartes ou les passions rêvées par la raison: Essai sur la théorie des passions de Descartes et de quelques-uns de ses contemporains.* Paris: Vrin, 2002.

Tarski, Alfred. "The Concept of Truth in Formalized Languages." In *Logic, Semantics, Metamathematics: Papers from 1923 to 1938*, ed. J. Corcoran, 152–278. Translated by J. H. Woodger. Indianapolis: Hackett, 1983.

Taylor, Sunaura. "Beasts of Burden: Disability Studies and Animal Rights." *Qui Parle* 19 (2011): 191–222.

Topsell, Edward. *The Historie of Serpents.* London: Jaggard, 1608.

Tort, Patrick, ed. *Darwinisme et société.* Paris: PUF, 1992.

Valdier, John. *L'exception humaine.* Paris: Éditions du Cerf, 2011.

Van Damme, Stephane. *Descartes: Essai d'histoire culturelle d'une grandeur philosophique.* Paris: Presses de Sciences Po, 2002.

Vicart, Marion. "Des chiens auprès des hommes: Ou comment penser la présence des animaux en sciences sociales." PhD diss., École des Hautes Études en sciences sociales, Paris, 2010.

Vilmer, Jean-Baptiste Jeangène. *Anthologie d'éthique animale: Apologies des bêtes.* Paris: PUF, 2011.

Vilmer, Jean-Baptiste Jeangène. *Ethique animale.* Preface by Peter Singer. Paris: PUF, 2008.

Vilmer, Jean-Baptiste Jeangène. "Les sophismes de la corrida." *Revue semestrielle de droit animalier* 2 (2010): 119–124.

Von Neumann, John, and Arthur Burks. *The Theory of Self-Reproducing Automata.* Champaign-Urbana: University of Illinois Press, 1966.

Von Uexküll, Jakob. *A Foray into the Worlds of Animals and Humans.* Translated by Joseph O'Neill. Minneapolis: University of Minnesota Press, 2010.

Warner, Michael. "Homo-narcissism; or, Heterosexuality." In *Engendering Men: The Question of Male Feminist Criticism*, ed. Joseph Boone and Michael Cadden, 190–206. New York: Routledge, 1990.

Whiteside, Kerry. *Divided Nature: French Contributions to Political Ecology.* Cambridge, MA: MIT Press, 2002.

Wittgenstein, Ludwig. *The Wittgenstein Reader.* Edited by Anthony Kenny. Oxford: Blackwell, 1994.

Wolch, Jennifer. "Zoöpolis." In *Animal Geographies: Place, Politics, and Identity in the Nature-Culture Borderlands*, ed. Jennifer Wolch and Jody Emel, 119–138. New York: Verso, 1998.

Wolfe, Cary. "Flesh and Finitude: Thinking Animals in (Post)Humanist Philosophy." In *The Political Animal*, ed. Chris Danta and Dimitris Vardoulakis, a special issue of *Substance* 37 (2008): 8–36.

Wolfe, Cary, ed. *Zoontologies: The Question of the Animal.* Minneapolis: University of Minnesota Press, 2003.

Wolff, Francis. *Philosophie de la corrida*. Paris: Fayard, 2007.

Wright, John. "The Embodied Soul in Seventeenth-Century French Medecine." *Canadian Bulletin of Medical History/Bulletin canadien d'histoire de la médecine* 8 (1991): 21–42.

# Contributors

**Éric Baratay** is a historian and professor at the Université de Lyon. A leading scholar in analyzing human/animal relationships from a historical perspective, Baratay is developing a different version of history, an animal history, by collaborating with ethologists in order to arrive at carefully recounted animal biographies (*Le point de vue animal*, Paris: Seuil, 2012).

**Nathalie Blanc** is the director of research in geography at the CNRS (Centre National de la Recherche Scientifique/École des Hautes Études en Sciences Sociales). Her research domains include the theme of nature in the city and environmental aesthetics.

**Florence Burgat** is a philosopher and director of research at the INRA (Institut National de la Recherche Agronomique) in Paris and a statutory member of the Archives Husserl de Paris. Her current work examines the condition of animals in industrial societies (*L'animal dans les pratiques de consommation*; *La protection de l'animal*) and explores phenomenological approaches to animal life (*Liberté et inquiétude de la vie animale*; *Une autre existence: La condition animale*). She has edited many collective volumes and is also coeditor of *Revue semestrielle de droit animalier*.

**Isabelle Delannoy** is an agricultural engineer specializing in bringing ecological issues to public awareness in the media. She is also a writer, coauthor of the eco-documentary *HOME* produced by Yann Arthus-Bertrand (2009), and cofounder of the ecological intelligence agency "Do Green" and the communication and strategy agency "Good Save the Green."

**Vinciane Despret** is Professor of Philosophy at the Université de Liège in Belgium. A philosopher and psychologist, she works at the forefront of animal studies as they come together in science, art, and literature. She has written extensively on animal representation in the arts (*Bêtes et hommes*), human-animal relationships as seen through her work with French farmers (*Être bête*), and philosophy as a way of understanding "le faire connaissance avec l'animal" (*Penser comme un rat*).

**Carla Freccero** is Professor of Literature, Feminist Studies, and History of Consciousness at the University of California at Santa Cruz, where she has taught since 1991. She also directs the UCSC Center for Cultural Studies. She is the author of *Queer/Early/Modern* (2006). Her book-in-progress, *Animate Figures*, explores literary figurations of animal being.

**Marcela Iacub** is a legal scholar at the CNRS and the École des Hautes Études en sciences sociales. She has published and taught widely on feminism, sexuality, and bioethics as they relate to French law and the right to individual choice concerning sexual behaviors and reproduction. Some of her many books include *Le crime était presque sexuel* (2002), *Par le trou de la serrure: Une histoire de la pudeur publique* (2008), and *Confessions d'une mangeuse de viande* (2011), in which she describes her conversion to vegetarianism.

**Dominique Lestel** is a philosopher currently living in Japan and doing research at the University of Tokyo. He is Professor of philosophy at the École Normale Supérieure (ENS) in Paris where he founded the Department of Cognitive Sciences (Département d'Etudes Cognitives). For a number of years, he has been developing a field-based philosophy that has contributed to an ethic and an ontology that are both relational and constructivist, linking the human and the other-than-human (animals, plants, robots, ghosts). He has published widely on a variety of topics: the meaning of the experiences of signing apes (*Paroles de singes*, 1995), the concept of human-animal friendship (*Les amis de mes amis*, 2007), the complexity of animal intelligences (*Les origines animales de la culture*, 2001), the epistemological limits of etholoy and comparative psychology (*L'animal est l'avenir de l'homme*, 2010), the irreducible singularity of animals as a species (*L'animal singulier*, 2004), the need to think about the human within the texture of animality and the importance of hybrid human-animal communities (*L'animalité*, 1996), and the ethical and moral issues of eating meat (*Apologie du carnivore,* 2011).

**Louisa Mackenzie** is Associate Professor in the Department of French and Italian Studies at the University of Washington in Seattle. Her research and publications focus mainly on the literature and culture of the French sixteenth-century, which she reads through various interpretive lenses that take into account the coproduction of natural and human worlds. Her book, *The Poetry of Place: Lyric, Landscape, and Ideology in Renaissance France* (2011), is an ecocritical study that situates lyric poetry in the landscapes and environments of sixteenth-century France. It won an honorable mention from the Modern Language Association in 2012. She has turned her attention to the role of animals in French humanist thought, and she has published articles on sea monsters and animal symbiosis, which will form part of a future book project.

**Marie-Hélène Parizeau** is Professor of Philosophy at the Université Laval in Quebec. Working in the fields of bioethics and environmental ethics, she has published numerous books on the idea of consent of the ethical subject as it pertains to both human and animal subjects (*L'être humain, l'animal et la technique; Néoracisme et dérives génétiques; Repenser la nature*). Her recent work on biogenetic engineering examines how hybrids push the limits of our current understanding of the human animal.

**Stephanie Posthumus** is an Assistant Professor in the Department of Languages, Literatures, and Cultures at McGill University, Montreal. Her area of specialization is twentieth- and twenty-first-century French literature, and she has published articles on representations of nature, landscapes, and animals as represented in works by Marie Darrieussecq, Michel Houellebecq, Jean-Christophe Rufin, and Michel Tournier. Working to develop a culturally specific French ecocriticism, she draws on the environmental philosophy of Michel Serres and the political ecology of Bruno Latour. Her doctoral thesis on nature and ecology in the work of Claude Lévi-Strauss, Michel Serres, and Michel Tournier was published in 2010. In this work, she traces the development of structuralist and ecologist models as they inform thinking about nature in anthropology, literature, and philosophy in the second half of the twentieth century in France. She is currently working on a book project on French *écocritique* that examines the cultural contexts in which environmental and animal issues are raised in contemporary French literature.

**Walter Putnam** is Professor of French and Chair of the Department of Foreign Languages & Literatures at the University of New Mexico and a Research Associate at the University of Johannesburg. His recent work has dealt with animals at the intersection of colonialism, on which he has numerous articles published and forthcoming, including "Cultural Displacements in Marie Nimier's La Girafe," in *Dalhousie French Studies*, and "The Colonial Animal," in *Animals and Society: An Introduction to Human-Animal Studies.*

**Peter Sahlins** is Professor of History at the University of California, Berkeley. A historian of France and Europe, he is considered a leading authority on the social and legal history of early modern nationality, citizenship, and boundaries. He is now working on a book that considers the uses of animals in multiple cultural contexts during the early reign of Louis XIV.

**Anne Simon** is a researcher at the CNRS in Paris. A leading literary scholar in the area of animal studies in France, she has been the principal investigator of two important research projects: "Animots: Animaux et animalité dans la littérature de langue française (XXe–XIXe siècles)" (2010–2014), and "Animalittérature" (2007–2010). She has published widely on the role of animals in twentieth-century French fiction, most notably in the work of Marcel Proust.

**Jean-Baptiste Jeangène Vilmer** holds degrees in three different disciplines: philosophy (BA, MA, PhD), law (LLB, LLM), and political science (PhD). He worked on animal ethics (*Ethique animale* (2008), *Philosophie animale* (2010), *L'éthique animale* (2011), and *Anthologie d'éthique animale* (2011), but his main research field is international relations, the laws and ethics of war in particular. Now a policy advisor on security issues at the French Ministry of Foreign Affairs and International Development, he teaches at Sciences Po Paris and the French military academy L'Ecole Spéciale Militaire de Saint-Cyr.

# Index

## B



## C

## D

## E

## M

## Z